Earth Crust

Edited by Muhammad Nawaz,
Farha Sattar and Sandeep Narayan Kundu

Published in London, United Kingdom

IntechOpen

Supporting open minds since 2005

Earth Crust
http://dx.doi.org/10.5772/intechopen.78850
Edited by Muhammad Nawaz, Farha Sattar and Sandeep Narayan Kundu

Contributors
Gencay Sariisik, Tongwei Zhang, Maman Hermana, Maya Genisa, Chow Weng Sum, Luluan Almanna Lubis, Boris Tarasov, Shaohua Yang, Yaolin Shi, Muhammad Nawaz

First published in London, United Kingdom, 2019 by IntechOpen
IntechOpen is the global imprint of INTECHOPEN LIMITED, registered in England and Wales, registration number: 11086078, 7th floor, 10 Lower Thames Street, London, EC3R 6AF, United Kingdom
Printed in Croatia

British Library Cataloguing-in-Publication Data
A catalogue record for this book is available from the British Library

Additional hard and PDF copies can be obtained from orders@intechopen.com

Earth Crust
Edited by Muhammad Nawaz, Farha Sattar and Sandeep Narayan Kundu
p. cm.
Print ISBN 978-1-78984-059-9
Online ISBN 978-1-78984-060-5
eBook (PDF) ISBN 978-1-83962-802-3

For EU product safety concerns:
IN TECH d.o.o., Prolaz Marije Krucifikse Kozulić 3, 51000 Rijeka, Croatia, info@intechopen.com or visit our website at intechopen.com.

Meet the editors

Dr. Nawaz is a geomorphologist with expertise in geographic information systems (GIS), spatial decision support systems, environmental science, and geospatial science teaching and learning. Since 1993, Dr. Nawaz has worked in different universities, including Punjab University in Pakistan, Charles Darwin University, Australia, and the National University of Singapore. He has vast experience and knowledge in the design, development, delivery, coordination, and teaching of courses in physical geography, GIS, environmental science, and geospatial science. Dr. Nawaz has more than 25 years of research over a broad range of issues relating to catchment management, sustainable development, sediment budget, and geospatial science teaching and learning.

Dr. Farha Sattar completed her Doctorate degree in Geospatial Science focusing on spatial mathematical modeling for 3D gully mapping and erosion quantification within a geoinformatics framework from the Charles Darwin University, Australia. Dr. Farha has more than 24 years' teaching experience in geographic information systems (GIS), spatial mathematics, and e-learning technologies at undergraduate and postgraduate levels. She has worked in several universities, including the University of the Punjab and the University of Twente, the Netherlands. Dr. S has been working at Charles Darwin University, Australia, since 2008. Prior to joining the School of Education, she worked as a lecturer in GIS in the School of Environment, Charles Darwin University, a lecturer in GIS in the GIS Centre, University of the Punjab, and a project manager in GIS/LIS in the National Database and Registration Authority, Pakistan. Dr. Farha is a certified drone pilot and aeronautical radio operator. She is well equipped with geospatial quantitative techniques and methods and has the experience to supervise research projects at postgraduate level. She has contributed to the curriculum development of GIS, remote sensing, and geography courses. Her research interests include drone technology, ICT in geospatial science, spatial mathematics education, GIS, and remote sensing.

Dr Sandeep Narayan Kundu is a geoscientist with more than 20 years of experience in industry and academia alike. He is currently responsible for designing geologically sound routes for continental submarine energy and telecommunication cables and pipelines at Fugro Singapore Marine Private Limited, and teaches geosciences as an adjunct faculty at the Department of Civil and Environmental Engineering, National University of Singapore. He has worked in various roles in the geology, geophysics, and petrophysics domain for the energy exploration industry at multinational organizations like BHP Billiton and Reliance Industries. His academic associations were with Friedrich Schiller University, Jena (Germany), Institute of Geoinformatics, University of Muenster

(Germany), Indian Institute of Technology (ISM, India), and the Indian Institute of Technology, Kanpur (India). He was awarded the prestigious British Council Fellowship and the Junior Research Fellowship in Earth Sciences and has delivered several keynote addresses at international forums. He is a member of SEAPEX and ICPC and has research interests in geospatial technology applications in geosciences and allied disciplines like environmental change, seafloor surveys, and geohazards.

Contents

Preface

The crust is the hard, outermost layer of our planet, made up of different types of rocks. Although it is a thin layer, and is even less than 1% of the total volume of the Earth, it is very important because of its ability to sustain life. Dynamic geomorphic and geological forces have created the crust and tectonic activity is constantly playing its role in transforming the crustal materials that underlie the ocean basins and others that form the continents.

The rocks that form the crust are mostly composed of silicon, oxygen, calcium, aluminum, potassium, sodium, magnesium, and iron. These rocks are distinctive in their chemical composition and different from the underlining ultramafic upper mantle rocks. The crust is part of the lithosphere, which also includes the hard and brittle rock of the uppermost part of the upper mantle.

This book aims to cover the basics of the architecture, structure, evolution, and dynamics of the Earth's crust through an anthology of contributed chapters that will enlighten readers about the various aspects of the Earth's crust, including the existence, development, and sustainability of our modern lifestyles on its surface.

Dr. Muhammad Nawaz
National University of Singapore,
Singapore

Chapter 1

Introductory Chapter: Earth Crust - Origin, Structure, Composition and Evolution

Muhammad Nawaz

1. Introduction

Earth crust is the thinnest and the most rudimentary layer that makes up the Earth, and yet, everything that has ever lived on Earth has called it home. The crust is a dynamic structure and it is one of the layers that make up our pale blue dot. The crust is referred to as a chemical layer that has varying chemical compositions. Two main types of crust are the oceanic crust and the continental crust, and they are different from each other. The differences are due to plate tectonics which then refers to plates and the movement of the plates above the asthenosphere, driving lithospheric processes that result in the formation and production of natural phenomena such as earthquakes and ridges. The crust is one of the five chemical layers of the Earth and is differentiated to show the distinct chemical properties occurring at each layer. The Earth's crust, along with the upper mantle, has its necessary role in the dynamic creation and destruction of the crustal surface in which all living organisms thrive on. This chapter will look at various aspects of the crust, and it will discuss the origin, structure and composition of the crust before elucidating its continued evolution to this day.

2. Origin

The early terrestrial crust appeared approximately 4.5 billion years ago, after the late stages of planetary accretion. This section describes the theories of the formation of the crust and discusses the origin of the oceanic and continental crust.

2.1 Theories about the formation of the crust

There are three main theories on the formation of the Earth's crust [1]: (1) inhomogeneous or heterogeneous accretion of the Earth model, (2) impact model and (3) terrestrial model.

The inhomogeneous model or the so-called the accretion model explains that the Earth's crust was formed during the accretion of the planet, with lighter and volatile elements forming a thin layer on the primitive planet which became the crust. This model suggests that non-volatile elements can only be found in the mantle; however, this is not true. Nonvolatile elements such as uranium and thorium are found on the Earth's crust [2], making this theory highly unlikely.

The impact model suggests that asteroids and other objects that impacted Earth melted and formed the crust [1]. The oceanic crust, which is mainly composed of basalt, could have been formed by a basalt asteroid that impacted the Earth. However, from the observations of the moon, basalts found in lunar maria were not due to an asteroid collision. Furthermore, the number of basalts produced from an impact event was too insignificant to form crusts [1]. In addition, a majority of the impact events on Earth happened after oceanic crusts were formed. Therefore, this theory is also unlikely as well.

The terrestrial model is the most likely explanation on the formation of the Earth's crust. This model explains that the crustal origin of the Earth was due to its internal processes. After the late accretion of the Earth, heat retained by the Earth resulted in the complete melting of the upper mantle, which formed a magma ocean that covered the surface of the Earth. As the Earth cooled, the magma ocean crystallised to form a widespread crust [1]. Another possible explanation was that the melted upper mantle rose up the surface to form a crust. The terrestrial model is the most likely explanation, as the magma ocean could explain some properties of the Earth's crust. The uniform composition of the crust could be formed by a homogeneous magma ocean. The layered composition of Earth's crust may be due to the cooling of magma oceans over time. Thus, the terrestrial model most likely explains the formation of the Earth's crust.

2.2 Origin of the oceanic crust

The oceanic crust was formed about 4.5 billion years ago, earlier than the first appearance of the continental crust, and it was first generated along the ocean ridges. The early oceanic crust differs from the present oceanic crust, in terms of its formation speed and thickness. The early crust was likely to be 20 km thick due to the high temperatures of the upper mantle. The higher mantle temperature caused a greater amount of melting in the upper mantle, resulting in more magma released to the surface to form thicker crusts [3]. The formation speed of the early oceanic crust was also likely to be faster than current speeds, due to the higher recycling rates caused by higher upper mantle temperatures [1]. The early oceanic crust is likely to be basalts in composition, and this could have resulted in the first plate tectonic activity. The basalt crust is denser than the molten mantle, so the basalt crust could have subsided into the upper mantle, leading to the recycling of crusts [3].

2.3 Origin of the continental crust

The oldest continental crust appeared about 4 billion years ago; however, granite continental crust only appeared about 3 billion years ago. There is no other planet in the solar system that has a continental crust except our Earth, mainly because it requires the presence of water on a planet and the subduction of crusts [4]. The seawater cools the hot mantle at the subduction zones, and it allows fractional crystallisation to take place to produce a granite crust [1].

3. Structure

The Earth has a thin silicate crust, which makes up 1% of the Earth's volume [5]. It is the uppermost top component of the lithosphere and floats on top of the upper mantle [6]. The crust plus the upper mantle is separated by the Mohorovicic discontinuity—a seismic and compositional boundary [6]. The crust varies in thickness as controlled by the law of isostasy according to Airy's model—the crust responds to topographical changes (loads or unloads) by changing its thickness as compensation, thus tending towards isostatic equilibrium [7]. The crust is thickest under mountain ranges and thinnest under mid-ocean ridges [6].

There are two main types of crust, the continental crust (underlie continents) and the oceanic crust (underlie ocean basins), the latter being denser and thinner but both being less dense than the mantle [6]. Approximately 35% of the Earth's crust is continental, while the other 65% is oceanic [8]. Continents are generally antipodic to oceans [9]. Conrad discontinuity, which lies at a depth of 5–20 km, separates the continental crust and oceanic crust [1]. Unlike continental crust, the oceanic crust has no granitic zone, but the mantle beneath the oceanic crust is possibly richer in radioactive elements than the mantle below the continental crust. The different locations of heat sources and thermal conductivity of the crust give rise to temperature variations in the different crust types.

Within the continental crust, there are four layers—the upper, middle, lower and the lowest layers [10]. The first layer is mainly made of sedimentary rocks and volcanic rocks, and the P-wave velocities in this layer are less than 5.7 km/s [10]. The second layer is mainly made of granitic plutons and metamorphic rocks (low-grade), and the P-wave velocities in this layer are between 5.7 and 6.4 km/s [11]. The third layer is mainly made of gabbroic cumulate, and the P-wave velocities in this layer range from 6.4 to 7.1 km/s [10]. The fourth layer is usually thin or missing [11], and the P-wave velocities in this layer are between 7.1 and 7.6 km/s [10].

The crust is carried as plates, which are slabs of lithosphere that carry oceanic crust, continental crust or both. They are carried by convection currents in the mantle, a process known as plate tectonics, which is driven by internal heat [1]. The plates meet at plate boundaries, which can be convergent boundaries, divergent boundaries or transform faults [1]. Interactions at plate boundaries, such as between crusts or between the crust and the mantle, can give rise to tectonic features such as oceanic ridges and volcanic arcs [6]. The crust undergoes physical and/or chemical changes in response to these interactions. For example, new oceanic crust develops at the opening rifts of divergent boundaries, forming mid-oceanic ridges through ridge push [12]. At convergent boundaries, dense oceanic crust subducts (slab pull) and produces magma due to the partial melting caused by the mantle's heat (thus creating a hotspot under the crust), which may rise and erupt, resulting in the formation of volcanic features, such as volcanoes, volcanic arcs and islands, on the non-subducted converging oceanic or continental crust [12]. This is seen in the Aleutian Islands. The colliding continental crust would result in the crust deforming into fold mountains [12], and an example of this is the Himalayan mountain range.

Continental margins are long narrow belts [13] that form at the outer edges of major landmasses [14] that include continental and submarine mountain chains. These margins could be passive, active or transform. Passive margins are found between continental and oceanic crust and are tectonically inactive, thus having a smooth relief [14]. Active margins have more tectonic and seismic activity and have features such as volcanoes and high sediment availability [14].

Another notable component of the Earth's structure is ophiolites. These are the fragments of oceanic crust and of the upper mantle that have undergone tectonic emplacement onto the continental crust [15]. They can be incorporated into both passive and active margins [16] and could be evidence of features of ancient oceanic crust that have since been consumed by subduction [15].

4. Composition

Minerals and rocks that make up the Earth's crust are the results of geological activity, density and tectonic plate movement. Minerals have definite chemical composition, whereas rocks are made up of minerals and have no specific chemical composition.

The three main kinds of rocks: igneous, sedimentary and metamorphic. Igneous rocks are formed by crystallisation of magma or lava. Sedimentary rocks are formed from lithification. Metamorphic rocks are formed from igneous and sedimentary rocks that undergone high temperatures and pressures, stress and fluid activity. The rock cycle ensures that these rocks are constantly replenished and recycled on the Earth's crust. Amongst the rocks, igneous rocks and metamorphic rocks make up 95% of the rocks [17], with granite and basalt having the largest compositions amongst the igneous rocks.

There are more than 3000 known minerals. Amongst them, only about 20 are common, and eight of these constitute 99% of the minerals in the crust. They are all silicates and are also called rock-forming minerals. Amongst the silicates, feldspars are the most abundant with plagioclase being the largest portion [18]. Minerals are formed by crystallisation through cooling of magma or lava and liquids. Another process is the evaporation of the liquid containing minerals, which result in the precipitation of material in the form of mineral veins.

4.1 Chemical composition

Elements are the building blocks of minerals. Oxygen and silicon are the most common elements in the crust (**Figure 1**). The Earth's crust consists of both oceanic crust and continental crust. Despite silicates being the most abundant minerals in both crusts, there are still some differences in their characteristics (**Table 1**). Thus, deriving the average compositions of each respective type of crust is essential and critical to investigate the continents and the Earth as such knowledge provides insights into the origin and characteristics of the crusts.

There are three primary methods used to study the Earth's composition: (1) studying and interpreting the seismic profile of both the core and mantle, (2) studying other planets and meteorites and comparing and inferring their composition and (3) using the pyrolite model [19]. Data from all complementary fields need to be integrated to allow a comprehensive and consistent understanding of the current dynamic structure and composition of the Earth.

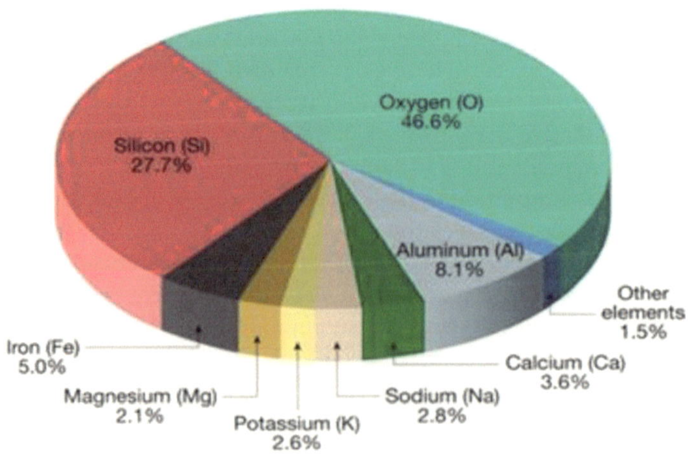

Figure 1.
Composition of elements of the Earth's crust.

Characteristics	Oceanic	Continental
Density	higher density (3.0 g/cm3)	more buoyant OC (2.6 g/cm3)
Thickness	7-10 km	25-70 km
Geological Age	Younger	Older
Chemical Composition	Basaltic (Sima)	Granitic (Sial)

Table 1.
Difference in characteristics.

4.2 Continental crust

Based on seismic investigations, the structure of continental crust is defined to consist of the upper crust, middle crust and lower crustal layers [10, 20]. Each layer varies slightly in its composition. The bulk composition is made mostly of rocks with a composition similar to granite rocks, full of substances such as oxygen, aluminium and silicon.

4.3 Oceanic crust

Similarly, oceanic crust is also layered, and each layer varies slightly in its composition [21]. In general, oceanic crust is basaltic and is rich in minerals and substances like silicon, oxygen and magnesium. To determine the chemical composition, it is important to look into mid-ocean ridge basalt (MORB). All the MORB reflects the mean composition of no or the zero-age ocean crust apart from back-arc basins [22].

5. Evolution

The evolution of the crust would refer to the gradual development of the crust over time. Geomorphically significant evolution of the Earth's crust falls into two main categories, endogenic processes from forces that originate within the Earth and exogenic processes that are a result of forces from above or on the planet surface.

5.1 Endogenous factors

Continental crust transforms into oceanic crust in a cyclic and dynamic process [23]. Where the old crust is being destroyed at convergent boundaries, new crust is being created at divergent boundaries. When rifting first occurs at divergent boundaries, the crust-mantle system transforms due to the temperature, and a rift forms. Subduction of the low-velocity zone in the upper part of the crust is the main mechanism overlooking the beginning of crustal attenuation. Intruding magma, originating from the mantle under the rift, modifies the intermediate and lower crustal layers. As the process continues, a "pseudo-oceanic" crust forms, which has an intermediate chemical composition. Before the new oceanic crust is created, the intermediate crust disappears completely, and the underneath crustal layer is critically modified by bouts of magma from the mantle sources. New oceanic crust is then produced from the ridge and spreads out from the spreading centre towards the subduction zone where the crust is eventually destroyed. Components of the crust will return to the upper crust in different forms such as igneous intrusions and contribute to the formation of new continental crust [21].Depending on the type of plate boundary and the types of plates involved, the resultant processes and landforms formed differ. The different phenomena that occur contribute to the evolution of the crust.

Another example of the evolution of the crust due to endogenous processes is volcanism, where material from the mantle or the deep crust is deposited onto the surface where it contributes in renewing the crust surface with new igneous rock and landforms. In some places the crust is weaker such as along plate boundaries, the magma forces its way through the rock, extruding rock and releasing pressure, which is why volcanic activity tends to occur near the borders of tectonic plates, for example, the Pacific Ring of Fire [22]. The composition and origin of the lava determine the type of volcanic landform created, with more fluid mafic lava forming structures such as shield volcanoes and more viscous felsic lava forming structures such as stratovolcanoes from the accumulation of ejecta. However, in cases where magma does not breach the surface, the magma in horns or magma chambers may solidify to form intrusive or plutonic rocks. Over time, the surrounding softer rock erodes away, revealing the harder plutonic rock beneath, which creates landforms such as plutons, batholiths, dykes, sills, laccoliths and volcanic necks.

5.2 Exogenous factors

The evolutionary processes mentioned above were all a result of forces originating from within the Earth. However, the crust is also shaped by a multitude of processes from external forces such as climate and extraterrestrial material. An overt example of an extraterrestrial force on the crust would be an impact crater, in which materials from space such as asteroids, meteoroids or comets collide with the Earth, leaving scars on the surface. While fairly infrequent in recent geological time, impacts were a major force of change during the late heavy bombardment period of the Earth's history [24], as the orbital path of the planet had not been fully cleared. The size of the impactor and extension diameter of the resultant impact crater is a decisive factor on the type of crater formed, with crater diameters above 2 km for sedimentary rocks and 4 km for crystalline rocks having a more complex impact structure as opposed to a simple bowl shape [25].

Climate and weathering are also significant drivers in the continued evolution of the crust. And while the parameters that control climate are complex and not fully understood, its effects can be seen widely. These processes can be observed in many forms, such as the exposure of batholiths by the erosion of soft rock, the carving of the Grand Canyon or the deposition of sediment by fluvial processes to create river deltas [26].

Additionally, biological processes also play a role in weathering and erosion. For example, plant roots hold the soil together, providing resistance to erosion [25]. Plants and burrowing animals also contribute to the mechanical breakdown of rock through wedging caused by growth and burrowing, respectively. All of the above processes are but a fraction of the factors that keep the Earth's crust in a state of constant flux. And while we may be unable to observe all geological evolutionary phenomena in the span of a human lifetime, we have more than enough examples and evidence to show that truly drastic changes occur in geological time.

6. Summary and conclusion

While the crust may only comprise the superficial layer of the Earth, it is truly a dynamic and fascinating thing to learn about. Superficially appearing to be a solid immutable covering of rock on our world, it is actually a collection of gargantuan rock plates of heterogeneous composition floating upon an equally colossal ocean of magma that is the outer mantle. From its early origins as a hot lifeless shell covering our planet to its current state as the home for all life on Earth, it has changed so much over the geological timespan of the Earth's history and continues to evolve to

DOI: http://dx.doi.org/10.5772/intechopen.88100

this day. It is fortunate that the Earth would coincidentally have the perfect chemical composition to form a crust suitable for life forms to exist, all in accordance with the physical laws that govern the formation of worlds. The crust is truly an amazing and astonishing thing to learn and behold.

Author details

Muhammad Nawaz
National University of Singapore, Singapore

*Address all correspondence to: geomn@nus.edu.sg

References

[1] Condie KC. Origin of the Earth's crust. Palaeogeography, Palaeoclimatology, Palaeoecology. 1989;**75**(1-2):57-81

[2] Heier KS. Radioactive elements in the continental crust. Nature. 1965;**208**(5009):479-480

[3] Van Thienen P, Van den Berg AP, Vlaar NJ. Production and recycling of oceanic crust in the early Earth. Tectonophysics. 2004;**386**(1-2):41-65

[4] Hargitai H, Kereszturi Á, editors. Encyclopedia of Planetary Landforms. New York: Springer; 2015

[5] Robertson EC. The interior of the Earth: An elementary description. Geological Survey Circular No. 532; 1996

[6] Volgyesi L, Moser M. The inner structure of the Earth. Periodica Polytechnica, Chemical Engineering. 1982;**26**:13-14

[7] Watts AB. Isostasy and Flexure of the Lithosphere. England: Cambridge University Press; 2001

[8] Cogley JG. Continental margins and the extent and number of the continents. Reviews of Geophysics. 1984;**22**(2):101-122

[9] Scheidegger AE. Principles of Geodynamics. New York: Springer Science & Business Media; 1963

[10] Rudnick RL, Fountain DM. Nature and composition of the continental crust: A lower crustal perspective. Reviews of Geophysics. 1995;**33**:267-309

[11] Yanagi T. Chemical composition of continental crust and the primitive mantle. In: Lecture Notes in Earth Sciences. Vol. 136. Tokyo: Springer; 2011. DOI: 10.1007/978-4-431-53996-4_2

[12] Wessel P, Müller RD. Plate tectonics. In: Treatise on Geophysics. Vol. 6. B.V.: Elsevier; 2015. pp. 45-93

[13] Ewing M, Press F. Structure of the earth's crust. In: Encyclopedia of Physics. Vol. 10/47. Berlin, Heidelberg: Springer; 1956

[14] Balasubramanian A, Mysore SKN. Continental Margins. 2016;**1**:1-7. DOI: 10.13140/RG.2.2.18278.22088

[15] Dilek Y, Furnes H. Ophiolites and their origins. Elements. 2014;**10**:93-100

[16] Zvi B. The emplacement of ophiolites by collision. Journal of Geophysical Research. 1982;**87**:16-18

[17] Shapley P. The Earth's Crust. Lecture Notes. University of Illinois. 2010. Available from: http://butane.chem.uiuc.edu/pshapley/Environmental/L26/1.html

[18] Crook T. The Earth's "Crust" and Its Composition. 1922. Retrieved from: https://www.nature.com/articles/110253a0

[19] McDonough W, Sun S-S. The composition of the Earth. Chemical Geology. 1995;**120**:223-253

[20] Holbrook WS, Mooney WD, Christensen NI. The seismic velocity structure of the deep continental crust. In: Fountain DM, Arculus R, Kay RW, editors. Continental Lower Crust. Amsterdam: Elsevier; 1992. pp. 1-44

[21] White W, Klein E. Composition of the Oceanic Crust. New York: Elsevier Ltd.; 2014

[22] Allaby A, Allaby M. A Dictionary of Geology and Earth Sciences. 4th ed. Oxford: Oxford University Press; 1999

[23] Mueller S. Evolution of the Earth's Crust. Dordrecht: D. Reidel Publishing Company; 1978

[24] Lowe DR, Byerly GR. The terrestrial record of late heavy bombardment. New Astronomy Reviews. 2018;**81**:39-61. DOI: 10.1016/j.newar.2018.03.002

[25] Huggett RJ. Fundamentals of Geomorphology. 4th ed. UK: Routledge Publishing; 2017

[26] Coleman S, Smart G. Fluvial sediment-transport processes and morphology. Journal of Hydrology. New Zealand. 2011;**50**(1):37-58. Retrieved from: http://www.jstor.org/stable/43945013

Chapter 2

Force and Specific Energy in Natural Rocks Cutting by Four-Axis Machine

Gencay Sariisik

Abstract

This study used a four-axis computer controlled machine to determine the cutting forces (C_f), specific cutting energy (S_c) and specific energy (S_e) affecting the processability of natural rocks. A total of 17 types of natural rocks were categorized according to their geological formations. This study consisted of three parts: (1) experimental measurement of the C_f resulting from processing natural rocks using a 6.0-mm diameter end mill cutting tool and a four-axis machine; (2) calculation of C_f, S_c and S_e values; and (3) statistical analysis of the parameters. In terms of C_f, S_c and S_e values, a statistically significant difference ($P < 0.001$) was observed among a depth of cut (d_p) and feed speed (V_a). It was found that the parameters that affected the processability of natural rocks were d_p and V_a. Accordingly, the processability of natural rocks processing type, d_p and V_a was found that affected. Simple linear regression was performed to examine the relationship between physicomechanical properties, and between C_f, S_e and S_c values. As a result, regression equations were developed in consideration of the physicomechanical properties of natural rocks and the C_f, S_c and S_e equation for the prediction of the efficiency of computer numerical control (CNC) machines.

Keywords: natural rock, CNC machine, efficiency, C_f, S_c, S_e

1. Introduction

Natural stone industry is one of the most effective actors of the Turkish mining industry economy. With its 4000-year history of natural stone production, Turkey is one of the oldest natural stone producers in the world. Natural stones are classified as metamorphic, sedimentary and magmatic according to their formations [1]. Marble is a metamorphic rock formed by recrystallization of limestone. Marble is a metamorphic stone composed of calcite ($CaCO_3$) and formed as a result of the recrystallization of the limestone under excessive pressure and heat [2–6]. CNC machines are electromechanical systems that process the materials according to the logical operation unit of numerical control (NC) codes defined by numbers, letters, and symbols of a product designed using a computer-aided design and manufacturing (CAD/CAM) program. Nowadays, 3D design in the natural stone industry can be processed with high accuracy, precision series and quality the use of CNC machine [7, 8]. Various processing technologies are used at natural stone quarries and factories to manufacture as building materials. In quarries, in the production of natural stone blocks, diamond wire cutting and chain arm cutting machines are widely used in factories, plates and panels manufacturing in the production of block cutter ST and gangsaw machines [9–18].

CNC machine is used to process natural stone blocks and wastes in 3D design products to be used in buildings. Many studies have been found on the C_f in circular saws depending on the physicomechanical properties of natural rocks, modeling of the S_e and socket [19–23], the definition of the theoretical chip geometry [24], and connections between tangential cutting force and chip thickness [25–27]. Some other studies address the effect of processing parameters on tool wear [28–31], the modeling of natural stone cutting with diamond cutting tools [32] and specific grinding energy of chip samples under a scanning microscope [33] in order to determine specific cutting energy and power consumption [34, 35]. Energy and the type of cutting mechanism used in the production of end products from natural rocks lead to the wear of cutting tools. Energy consumption and wear of cutting edges are the important factors affecting processing costs. These factors should be efficiently to reduce processing costs. It is therefore important to determine the cutting performance of natural rocks according to their characteristics. Cut tools should be selected carefully and performance forecast should be proper in order to improve the processing efficiency of natural rocks and reduce costs. Different from the manufacture of plates and panels in which gangsaw and cutter ST are deep cut on the x (length), z (depth) axis, this process requires few depth cut and precision tools with a CNC machine on the x (length), y (width) and z (depth) axis. C_f, S_c and wear of diamond cut tools have a low spindle and V_a in the processability of rocks in the CNC machine with internally cooled cut tools. Carbide coated end mills tools are widely used in the natural stones sectors. In addition, the number of studies on parameters affecting the processability of natural rocks in the CNC machine is less [36–40].

This study investigated the relationship between the values of the C_f, S_e and S_c obtained from the processing of rocks using CNC, and related parameters (i.e., cutting parameters and physicomechanical properties). C_f, S_e and S_c values depending on the d_p and V_a of natural rocks have been analyzed (ANOVA) statistically. Results show that 1.0, 1.2 and 1.6 mm at the d_p and 2000, 2500, 3000 mm/min at the V_a are important cutting parameters in the processability of natural rocks. Regression models were developed for the estimation of C_f, S_e and S_c of the with measured by power and load meter tester. Regression analyses were performed to determine the optimum relationship between the C_f, S_e and S_c dependent variable and physicomechanical properties independent variable in three different groups of natural rocks. Regression models developed in this study can be used by planners and natural stone manufacturing companies for cost analysis and mill cut tool processing programs in natural rocks. We recommend that it be considered in further studies use a larger database to analyze the possibility of regression analyses the C_f, S_e and S_c any of the physico-mechanical parameters of natural rocks.

2. Materials and method

2.1 Materials

These natural rock samples were obtained from the companies operating a quarry in Afyon. The surface of each series was adjusted with the cutting process as a square (30 by 30-cm) and 1 cm thickness. Furthermore, these rock samples were polished on one side. **Table 1** shows sample code, pertographic name, dimensions, main modal compositions, sample number and surface processing of the rocks used in the test.

Natural rocks	Sample code	Pertographic name	Main modal compositions	Dimensions (mm)	N	Surface processing
Travertine	T1-T5	Sedimentary	Calcite	300 × 300 × 30	15	Polished
Marble	M1-M7	Metamorphic	Calcite	300 × 300 × 30	15	Polished
Limestone	K1-K5	Sedimentary	Calcite	300 × 300 × 30	15	Polished

Table 1.
Characteristics of natural rocks used in experimental tests.

Cut tool (carbide-coated end-mill cutting edge tool) was used in the processing of natural rocks by CNC machine. **Figure 1** shows the image of the end mill cut tool. **Table 2** shows the technical specifications of the end mill cut tool.

2.2 Method

Figure 2 shows that the CNC consists of the machine, electric motors, pneumatic or hydraulic power units, control panel and software.

A 4-Axis CNC (Megatron) machine designed for the natural stone industry in Afyon Kocatepe University Natural Stone Processing Laboratory was used for the tests. The technical features of the machine are given in **Table 3**.

Processability tests were performed on natural rocks according to the d_p and V_a cutting parameters by using CNC machine. Alpha CAM drawing program was used to determine the cutting parameters of natural rocks. As shown in **Figure 3**, 120 × 25 mm in size 18 rectangular samples were produced using a 3D modeling and simulation program for the processability tests.

A power meter, also known as load meter, built in the CNC machine was used in the processability tests. This testing tool consists of a measurement unit, a load cell, a controller unit and Defne Lab Soft program (**Figure 4**).

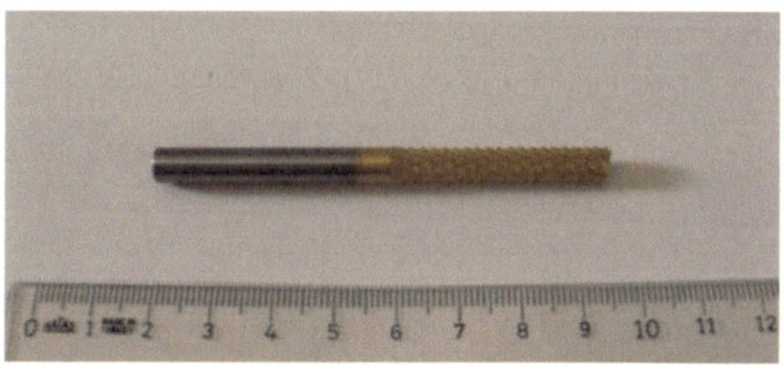

Figure 1.
Image of the end mill cutting edge tool.

Technical properties	Unit	Values
Code	—	MFR-6
Cutting tool diameter/d1	mm	6
Stem diameter/d2	mm	6
Cutting length/l2	mm	25
End length/l1	mm	76
Milling end	—	36,677
Edge number	—	4
Helix angle	(°)	25

Table 2.
Technical properties of end-mill cutting edge tool.

Figure 2.
CNC natural stone processing machines.

Technical properties/unit	Values
Spindle motor (kW)	9.0
Number of axis (number)	4.0
Motor speed (rpm)	24.000
Processing speed (rpm)	24.000
Vamotor x axis (mm/dk)	80.000
Voltage (V)	380
Processing length (mm)	4.000–4.500
Processing width (mm)	2.000–2.500
Processing height (mm)	500–600
Lathe height (mm)	700–750
Lathe length (mm)	2.500–3.000
Coolant (l/dk)	3.0
Automatic number of teams (adet)	8.0

Table 3.
Technical properties of CNC natural stone processing machine.

Natural stone samples were bond and stabilized to the platform on the measurement unit with cut tools. There were 4 on the Z axis, 4 on the X and Y axis total 8 load cells on the testing machine to calculate the C_f. Defne Lab-Soft program was recorded into the data input screen of types and sizes of natural rocks, cutter information, and constant and variable parameters. NC codes from Alphacam drawing program were transferred to the CNC machine with Recon program interface. The cut tool was balanced to the engine of the CNC machine connected to the tool holder. The reset operation was performed when the cutting tool was guided by the function and operation keys to the reference point on the test specimen surface. NC code was selected using the function and operation keys on the control unit and the measurement was performed by pressing the start button. Depending on the different process parameters, the natural stones were processed in the cooling process from the water flow rate of 1 l/min. A rectangle of 120 × 25 mm was processed for 40 s to obtain 100 data per second. All samples were processed in a total of 84 min. **Figure 4** shows a schematic view of the test apparatus.

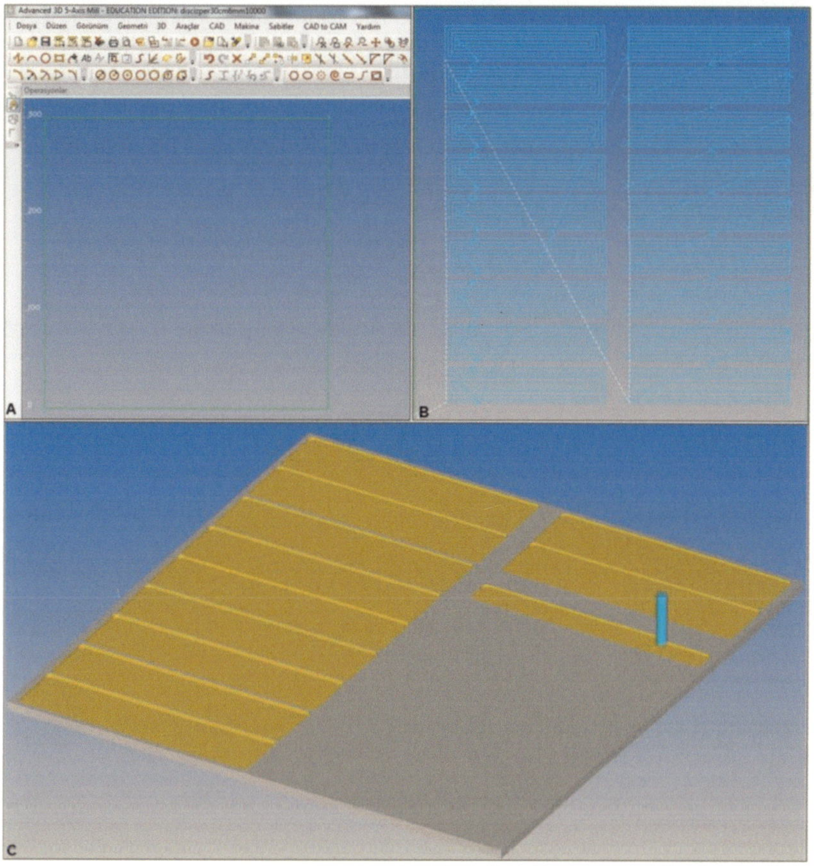

Figure 3.
Test specimens on (a) alpha CAM drawing program, (b) modeling, (c) simulation.

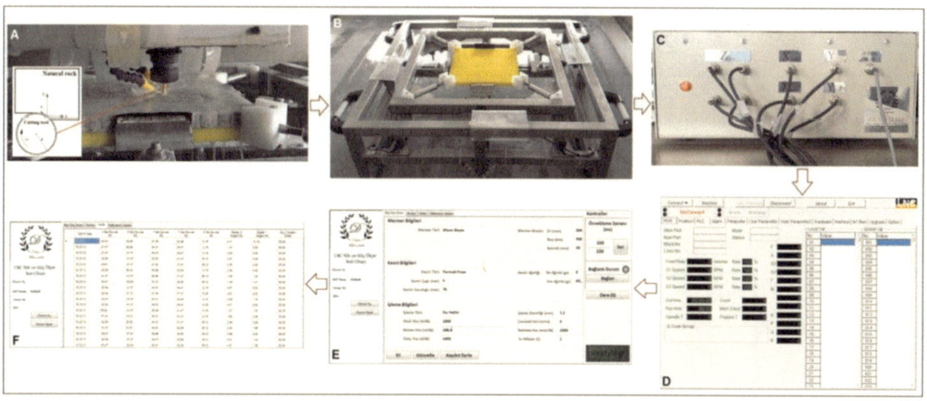

Figure 4.
Processability tests (a) vector representation of forces (F_x, F_y, F_c and F_t), cutting speed (V_t) and Va(V_a), (b) power and load meter tester and load cells, (c) control unit, (d) Defne lab-soft natural stone test program interface, (e) recon program interface, (f) views of database.

In the processability tests, variable and constant parameters were taken into consideration. Constant cutting parameters were the cut tool diameter of 6.0 mm, spindle speed of 10,000 d/min, cutting width of 3.0 mm and plunge speed of 1000 d/min. Variable cutting parameters were the d_p of 1.20, 1.60 and 2.0 mm and Va of

2500, 3000, 3500 mm/min. **Table 4** shows the CNC cutting parameters for the natural rocks processed in the tests.

Figure 5 explains the vector representation of the forces (F_x, F_y, F_c and F_t), V_t and V_a that occurred during the processing of the natural rocks.

Calculation of F_x cutting force according to CNC processing parameters is as shown in Eq. (1).

F_x cutting force Eq. (1);

$$F_x = |F_{x1}| + |F_{x2}| \tag{1}$$

F_x = cutting force (N); F_{x1} = absolute forward cutting force (N); F_{x2} = absolute back cutting force (N).

F_y cutting force Eq. (2);

$$F_y = |F_{y1}| + |F_{y2}| \tag{2}$$

F_y = cutting force (N); F_{y1} = absolute forward cutting force (N); F_{y2} = absolute back cutting force (N).

Calculation by using R resultant force, F_x and F_y cutting forces, Eq. (3):

$$R = \sqrt{F_x^2 + F_y^2} \tag{3}$$

R = resultant force (N); F_x = cutting force (N); F_y = cutting force (N).

β angle between R and Fx, Eq. (4),

Processing parameters	Unit	Values
Cutting tool diameter	mm	6.0
Depth of cut	mm	1.20–1.60–2.00
Spindle speed	d/min	10,000
Feed speed	mm/min	2000–2500–3000
Plunge speed	d/min	1000
Cutting speed	m/min	188.4
Cutting width	mm	3.0

Table 4.
CNC processing parameters.

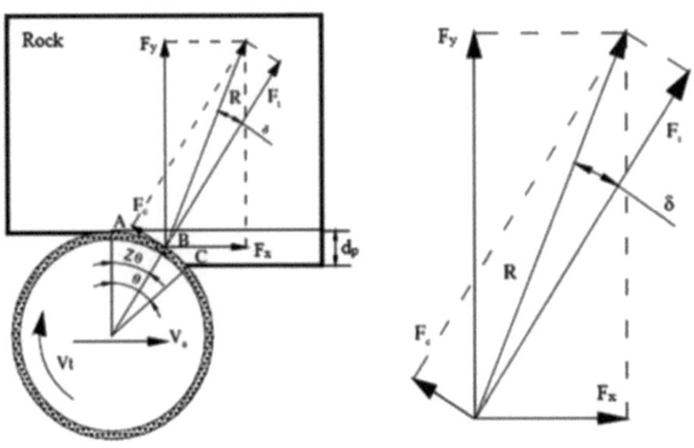

Figure 5.
Vector representation of the forces (F_x, F_y, F_c and F_t), V_t and V_a that occur during the processing.

$$\beta = tan^{-1}\left(\frac{F_y}{F_x}\right) \tag{4}$$

Contact angle θ between tool diameter (d) and natural rocks, Eq. (5);

$$\theta = cos^{-1}\left(1 - \frac{2dp}{d}\right) \tag{5}$$

Calculating F_c tangential force and radial force F_t components of the cutting forces with the R value obtained.

Eqs. (6) and (7):

$$F_c = Rsin\delta \tag{6}$$

$$F_t = Rcos\delta \tag{7}$$

δ angle between F_t and F_c.

Eq. (8):

$$\delta = \beta - Z\theta \tag{8}$$

Parameter Z depends on the location of the application point of the compound forces R on the arc AC, which is contact between cutting edges and natural rocks.

Parameter Z, Eq. (9):

$$Z - \frac{AB}{AC} \tag{9}$$

V_t cutting speed Eq. (10):

$$V_t = \frac{\pi \times D \times n}{1000} \tag{10}$$

V_t = cutting speed (m/min); n = spindle speed (d/min); D = cutter diameter (mm).

Specific cutting energy depending on tangential force and cutting speed is shown in Eq. (11).

$$S_c = \frac{F_c \times V_t}{V_a \times d_p \times b} \tag{11}$$

F_c = tangential cutting force (N); V_t = cutting speed (m/min); V_a = feed speed (mm/min); d_p = cutting depth (mm); b = cutting width (mm).

S_e values were calculated using the power P and Q_w obtained from the main electric motor of 7.5 kW where the cutting end of the natural rocks was connected during the processability time (t).

The chip volume is shown in Eq. (12).

$$Q_w = b \times l \times dp(1, 2, 3) \tag{12}$$

Q_w = chip volume (mm^3); b = size of the sample (mm); l = the width of the sample (mm); d_p(1,2,3) = cutting depth (mm).

The total specific energy is shown in Eq. (13).

$$S_e = \frac{\frac{\sum_{j=1}^{n} \mathrm{Pj}}{n} \times \sum_{j=1}^{n} \mathrm{tj}}{Q_{w(1,2,3)}} \tag{13}$$

S_e = total specific energy (J/mm^3); P = power consumption (W); t = total time (s); Q_w = chip volume (mm^3).

3. Findings and evaluations

3.1 Petrographic, chemical and physicomechanical properties

The petrographic analysis was carried out in the MTA (General Directorate Mineral Research and Exploration) mineralogical and petrographic analysis laboratory in Ankara/Turkey. Chemical properties of natural rocks were performed using the XRF (X-ray fluorescence) method in ACME (Analytical Laboratory in Turkey/ Ankara). The metamorphic (marble) and sedimentary (travertine, and limestone) origin natural rocks used in this study had different textures. All natural rock types were composed of CaO as main calcite crystals and at least 99.0% calcite minerals ranging from 53.10 to 55.70%. The petrographic analysis results and chemical analysis of the samples are presented in **Tables 5** and **6**, respectively. Petrographic and chemical analysis results of samples are given in **Tables 5** and **6**, respectively.

The CNC processability tests were conducted in the Rock Mechanics and Technology Application and Research Center Laboratory of the Department of Mining Engineering of Afyon Kocatepe University. The tests were performed in accordance with Standard No. TS EN 1936: 2010 [41], Standard No. TS EN 13755: 2014 [42], Standard No. TS EN 14205: 2004 [43], Standard No. TS EN 1926: 2007 [44], Standard No. TS EN 13161: 2014 [45], Standard No. TS 699 [46] and Standard No. TS EN 1341 (Appendix-C: 2013) [47]. The physicomechanical properties of the natural rocks are presented in **Table 7**. The rock samples were 40×40 mm^3, $70 \times 70 \times 70$ mm^3 and $30 \times 50 \times 180$ mm^3. The tests were carried out using at least six samples.

3.2 Variance analysis (ANOVA) of cutting forces

In processability tests, the F_c and F_t measurements were conducted using two-factor analysis of variance (ANOVA) (17 natural rocks $\times$ 2 C_f $\times$ 3 d_p $\times$ 3 V_a) randomized experimental design with 100 replications (n = 100). A total of 30,600 data were obtained on the rocks. In terms of the C_f (F_c, F_t), among the d_p and V_a there was a statistically significant difference ($P < 0.001$) (**Table 8**).

In processability tests for natural rocks, the mean F_c and F_t increase with an increase in the d_p and V_a was given in **Figure 6**. K4 and K5 samples have high values of the C_f at the d_p of 2.0 mm while T1, T2, and T3 samples have low values of the C_f at the d_p of 1.2 mm. Processability of the C_f values of K4 and K5 samples the d_p of 2.0 mm is more forced than that of the other samples. K4 and K5 samples have high values of the C_f at a V_a of 3.000 mm/min while T1, T2, and T3 samples have low values of the C_f at a V_a of 2.000 mm/min. Processability of the C_f of K4 and K5 samples at a V_a of 3.000 mm/min is more forced than that of the other samples.

3.3 Variance analysis (ANOVA) of specific cutting energy and specific energy

In processability tests, the S_c and S_e measurements were conducted using two-factor analysis of variance (ANOVA) (S_c and S_e for 12 natural rocks $\times$ 3 d_p $\times$ 3 V_a)

Natural Rocks	Petrographic descriptions	Minerals
T1	Fine-grained calcite is the dominant mineral. Consist of micro-mesocrystalline calcite minerals with a little amount of clay. Often contains pores. Micritic (intraclast) texture. Travertines	77% Calcite (mic), 22% Calcite (spr)
T2		78% Calcite (mic), 21% Calcite (spr)
T3		79% Calcite (mic), 20% Calcite (spr)
T4		78% Calcite (mic), 21% Calcite (spr)
T5		77% Calcite (mic), 22% Calcite (spr)
M1	Fine, medium, medium-coarse and coarse-grained with polysynthetic twins, granoblastic texture. Marbles	98.5% Calcite
M2		97.5% Calcite
M3		98% Calcite
M4		98.5% Calcite
M5		98.5% Calcite
M6		99% Calcite
M7		97% Calcite, 2% dolomite
K1	Fine-grained. Consists of cryptocrystalline calcite within crypto microcrystalline calcite. Micritic texture. Limestones	96% Calcite (mic), 2% Calcite (spr)
K2		96% Calcite (mic), 2% Calcite (spr)
K3		95% Calcite (mic), 3% Calcite (spr)
K4		95% Calcite (mic), 3% Calcite (spr)
K5		95% Calcite (mic), 3% Calcite (spr)

Table 5.
Petrographic descriptions of natural rock samples.

randomized experimental design with 100 replications (n = 100). A total of 30,600 data were obtained on the rocks. In terms of the S_c and S_e, among the d_p and V_a there was a statistically significant difference ($P < 0.001$) (**Table 9**).

In processability tests for natural rocks, the mean S_c and S_e values at the d_p of 1.2 mm are lower than those at the d_p of 1.6 and 2.0 mm was given in **Figure 7**. S_c and S_e values at the V_a of 2000 mm/min are higher than those at the V_a of 2500 and 3000 mm/min. S_c and S_e values at the V_a of 3000 mm/min are lower for T1, T2, and T3 samples while those at the V_a of 2000 mm/min are higher for both K4 and K5 samples. The natural rocks should have the V_a of 3000 mm/min according to the S_c and S_e values.

3.4 Relationships between cutting forces and specific cutting energy

Regression models were applied to examine the relationship between the C_f and S_c values for each of the natural rocks. The results of the simple linear regression analysis are given in **Figure 8**.

Natural Rocks	CaO (%)	SiO_2 (%)	Al_2O_3 (%)	Fe_2O_3 (%)	MgO (%)	K_2O (%)	TiO_2 (%)	P_2O_5 (%)	MnO (%)	LoI (%)
T1	55.70	0.01	0.01	0.01	0.01	0.015	0.01	0.03	0.02	43.96
T2	55.44	0.06	0.01	0.02	0.06	0.11	0.01	0.02	0.03	44.02
T3	55.45	0.01	0.01	0.01	0.28	0.12	0.01	0.02	0.01	44.08
T4	55.47	0.01	0.01	0.01	0.21	0.10	0.01	0.02	0.01	44.01
T5	55.48	0.06	0.01	0.02	0.09	0.11	0.01	0.02	0.03	43.95
M1	55.59	0.01	0.01	0.01	0.11	0.21	0.01	0.04	0.01	44.00
M2	54.40	0.95	0.31	0.05	0.21	0.04	0.01	0.01	0.01	44.00
M3	54.16	0.89	0.27	0.09	0.81	0.08	0.02	0.01	0.04	43.60
M4	53.37	1.57	0.57	0.29	0.75	0.27	0.03	0.03	0.04	43.06
M5	55.38	0.17	0.03	0.09	0.31	0.21	0.01	0.02	0.01	43.76
M6	55.08	0.27	0.03	0.19	0.33	0.19	0.01	0.02	0.01	43.86
M7	53.10	0.55	0.30	0.19	1.92	0.05	0.01	0.03	0.03	43.81
K1	55.57	0.10	0.02	0.01	0.37	0.16	0.01	0.04	0.01	43.71
K2	54.92	0.10	0.01	0.01	0.41	0.13	0.01	0.04	0.01	44.36
K3	54.80	0.22	0.01	0.01	0.40	0.11	0.01	0.04	0.01	44.39
K4	54.68	0.46	0.01	0.01	0.56	0.12	0.01	0.03	0.01	44.11
K5	54.37	0.75	0.01	0.01	0.35	0.14	0.01	0.04	0.01	44.31

Table 6.
Chemical characteristics of natural rock samples.

Natural Rocks	D (kg/m^3)	P (%)	WA (%)	KH	UCS (MPa)	FS (MPa)	IS (MPa)	AR ($cm^3/50\ cm^2$)
T1	2640	1.52	1.38	113.38	54.56	7.92	17.00	24.01
T2	2650	1.46	1.24	117.96	55.25	8.09	19.00	23.48
T3	2660	1.28	1.06	122.54	56.85	8.46	21.00	23.05
T4	2670	1.15	0.94	126.28	57.94	8.89	22.00	22.61
T5	2680	1.04	0.87	129.35	58.65	8.95	23.00	22.04
M1	2690	0.96	0.82	131.14	59.85	9.03	24.00	21.79
M2	2700	0.88	0.78	135.76	62.16	9.28	25.00	19.95
M3	2705	0.75	0.66	137.89	64.85	9.46	26.00	19.01
M4	2710	0.71	0.55	141.52	66.85	9.86	27.00	18.72
M5	2715	0.68	0.43	143.78	68.75	9.98	28.00	18.13
M6	2720	0.66	0.38	145.78	70.45	10.12	29.00	17.79
M7	2730	0.52	0.32	155.45	77.04	10.94	31.00	17.58
K1	2735	0.38	0.28	168.25	84.10	12.85	33.00	17.29
K2	2745	0.25	0.21	173.56	87.25	13.94	34.00	16.57
K3	2755	0.23	0.18	178.24	90.84	14.96	36.00	16.05
K4	2775	0.20	0.16	184.58	93.26	16.24	38.00	15.34
K5	2800	0.16	0.14	190.95	97.08	18.45	40.00	14.12

D, density; P, porosity; WA, water absorption; KH, knoop hardness; UCS, uniaxial compressive strength; FS, flexural strength; IS, impact strength; AR, abrasion strength.

Table 7.
Physico-mechanical properties of natural rocks.

Cf (N) dependent variable	d_p (mm)		Mean difference (I-J)	Std. Error	Sig.	95% confidence interval	
	Mean (I)	Mean (J)				Lower bound	Upper bound
F_c	1.20	1.00	−3.8214*	0.27897	<0.001	−4.4787	−3.1642
		2.00	−7.4694*	0.27897	<0.001	−81267	−6.8122
	1.60	1.20	3.8214*	0.27897	<0.001	3.1642	4.4787
		2.00	−3.6480*	0.27897	<0.001	−4.3052	−2.9907
	2.00	1.20	7.4694*	0.27897	<0.001	6.8122	8.1267
		1.60	3.6480*	0.27897	<0.001	2.9907	4.3052
F_t	1.20	1.60	−3.7896*	0.27690	<0.001	−4.4420	−3.1372
		2.00	−7.5726*	0.27690	<0.001	−8.2250	−6.9202
	1.60	1.20	3.7896*	0.27690	<0.001	3.1372	4.4420
		2.00	−3.7830*	0.27690	<0.001	−4.4354	−3.1306
	2.00	1.20	7.5726*	0.27690	<0.001	6.9202	8.2250
		1.60	3.7830*	0.27690	<0.001	3.1306	4.4354
Va(mm/dk)							
F_c	2000	2500	−2.2064*	0.27897	<0.001	−2.8637	−1.5491
		3000	−3.7687*	0.27897	<0.001	−4.4259	−3.1114
	2500	2000	2.2064*	0.27897	<0.001	1.5491	2.8637
		3000	−1.5623*	0.27897	<0.001	−2.2195	−0.9050
	3000	2000	3.7687*	0.27897	<0.001	3.1114	4.4259
		2500	1.5623*	0.27897	<0.001	0.9050	2.2195
F_t	2000	2500	−2.1094*	0.27690	<0.001	−2.7618	−1.4570
		3000	−3.3713*	0.27690	<0.001	−4.0236	−2.7189
	2500	2000	2.1094*	0.27690	<0.001	1.4570	2.7618
		3000	−1.2619*	0.27690	<0.001	−1.9142	−0.6095
	3000	2000	3.3713*	0.27690	<0.001	2.7189	4.0236
		2500	1.2619*	0.27690	<0.001	0.6095	19142

The mean difference is significant at the, 05 level.

Table 8.
Statistical analysis of C_f of natural rocks depending on the d_p and V_a.

Figure 8 shows that there is a statistically significant relationship between S_c and C_f values. Correlation coefficient (R^2) values obtained from the natural rocks in the F_c at depths of cut of 1.2, 1.6 and 2.0 mm are 0.895, 0.871 and 0.859, respectively. Correlation coefficient (R^2) values obtained from the natural rocks in the F_t at depths of cut of 1.2, 1.6 and 2.0 mm are 0.890, 0.878 and 0.880, respectively. Correlation coefficient (R^2) values obtained from the natural rocks in the F_c at the V_a of 2000, 2500 and 3000 mm/min are 0.771, 0.780 and 0.780, respectively. Correlation coefficient (R^2) values obtained from the natural rocks in the F_t at the V_a of 2000, 2500 and 3000 mm/min are 0.745, 0.781 and 0.781, respectively.

3.5 Relationships between specific cutting energy and specific energy

The results of the correlation coefficient and regression model are given in **Figure 9**. Correlation coefficient (R^2) obtained from the natural rocks is 0.784, indicating that there is a linear relationship between the S_c and S_e.

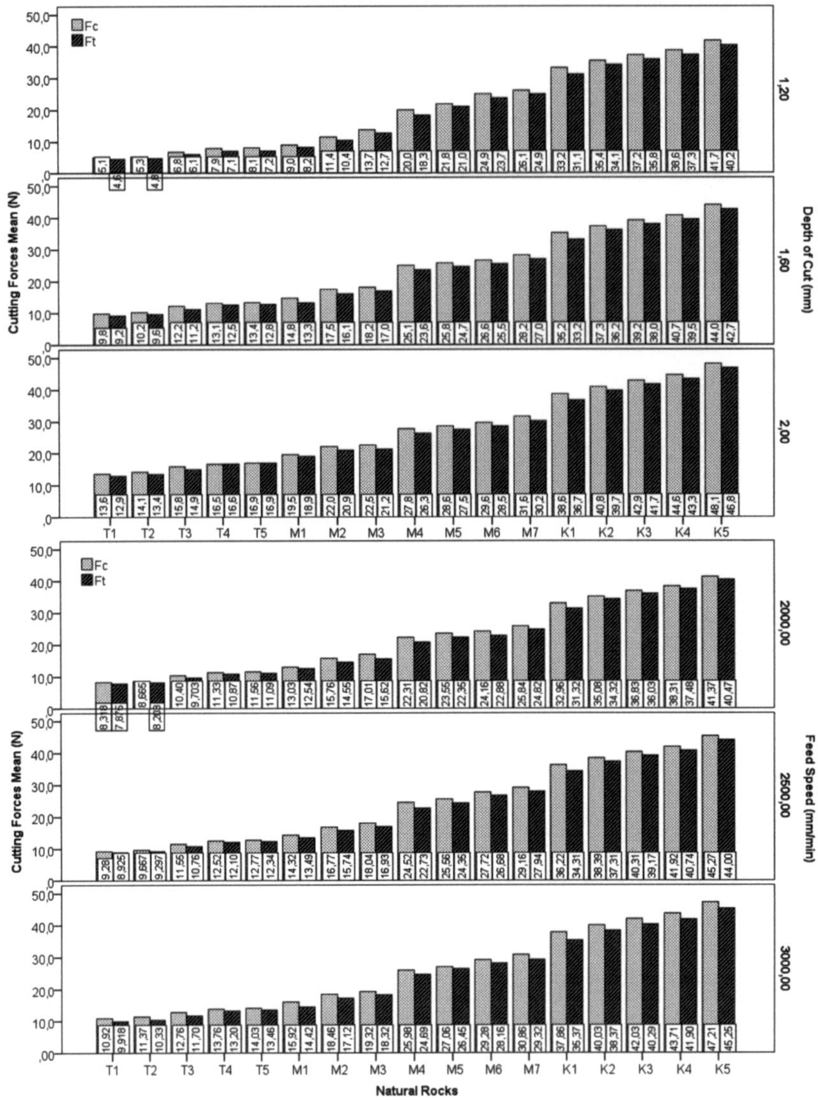

Figure 6.
C_f according to the d_p and V_a in natural rocks.

3.6 Relationships between cutting forces, specific cutting energy and specific energy natural rocks properties

The relation between the F_c, F_t, S_c and S_e values, and the physico-mechanical properties of the natural stones were evaluated through regression analysis. The results of the regression analysis are given in **Tables 10** and **11**, respectively.

Correlation coefficient (R^2) values of the natural rock samples, a linear relationship between the C_f values and the physicomechanical characteristics is observed. As a result of the analysis, the following correlation coefficient (R^2) values have been obtained: R^2 coefficient range from 0.887 to 0.981 in the F_c and from 0.883 to 0.983 in the F_t. All values confirm the linear relationship among physicomechanical properties in natural rocks with the C_f. Accordingly, as porosity, water absorption and abrasion strength decrease in the natural rocks, C_f values increases. Moreover,

S_c and S_e dependent variable	d_p (mm)		Mean difference (I-J)	Std. Error	Sig.	95% confidence interval	
	Mean (I)	Mean (J)				Lower bound	Upper bound
S_c	1.20	1.60	72898*	0.70556	<0.001	56.275	89.521
		2.00	10.9091*	0.70556	<0.001	92.468	12.5714
	1.60	1.20	−72898*	0.70556	<0.001	−89.521	−56.275
		2.00	36193*	0.70556	<0.001	19.569	52.816
	2.00	1.20	−10.9091*	0.70556	<0.001	−12.5714	−92.468
		1.60	−36193*	0.70556	<0.001	−52.816	−19.569
V_a(mm/dk)							
S_c	2000	2500	45448*	0.70556	<0.001	28.825	62.071
		3000	95051*	0.70556	<0.001	78.427	11.1674
	2500	2000	−45448*	0.70556	<0.001	−62.071	−28.825
		3000	49603*	0.70556	<0.001	32.980	66.226
	3000	2000	−95051*	0.70556	<0.001	−11.1674	−78.427
		2500	−49603*	0.70556	<0.001	−66.226	−32.980
S_e	1.20	1.60	14500*	0.02103	<0.001	14.004	14.996
		2.00	23077*	0.02103	<0.001	22.582	23.573
	1.60	1.20	−14500*	0.02103	<0.001	−14.996	−14.004
		2.00	.8577*	0.02103	<0.001	.8082	0.9073
	2.00	1.20	−23077*	0.02103	<0.001	−23.573	−22.582
		1.60	−0.8577*	0.02103	<0.001	−0.9073	−0.8082
V_a(mm/dk)							
S_e	2000	2500	0.6747*	0.02103	<0.001	0.6252	0.7243
		3000	11701*	0.02103	<0.001	11.205	12.197
	2500	2000	−0.6747*	0.02103	<0.001	−0.7243	−0.6252
		3000	0.4954*	0.02103	<0.001	0.4458	0.5449
	3000	2000	−11701*	0.02103	<0.001	−12.197	−11.205
		2500	−0.4954*	0.02103	<0.001	0.5449	−0.4458

**The mean difference is significant at the, 05 level.*

Table 9.
Statistical analysis of S_c and S_e of natural rocks depending on the d_p and V_a.

as knoop hardness, uniaxial compressive strength, flexural strength and impact strength increase, the C_f values also increase.

Correlation coefficient (R^2) values of the natural rock samples, a linear relationship between the S_c and S_e values and the physicomechanical characteristics is observed. As a result of the analysis, the following correlation coefficient (R^2) values have been obtained: R^2 coefficient range from 0.887 to 0.977 in S_c and from 0.616 to 0.858 in the S_e. All values confirm the linear relationship among physicomechanical characteristics in natural rocks with the S_c and S_e. Accordingly, as porosity, water absorption and abrasion strength decrease in the natural rocks, S_c and S_e value increases. Moreover, as Knoop hardness, uniaxial compressive strength, flexural strength and impact strength increase, the S_c and S_e values also increases.

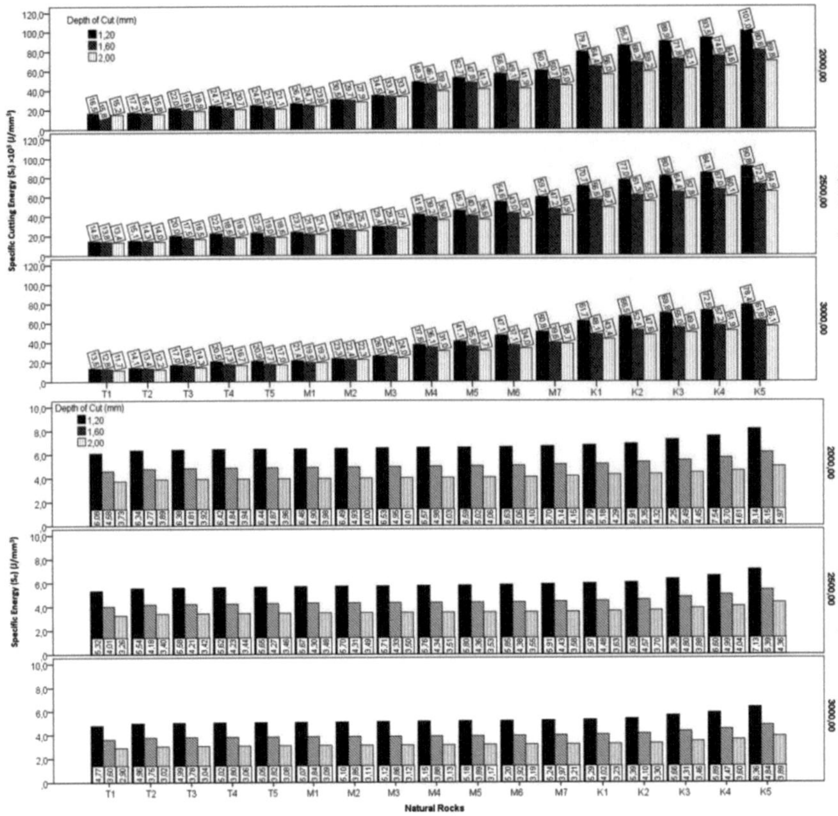

Figure 7.
S_c and S_e values according to the d_p in processability tests of rocks.

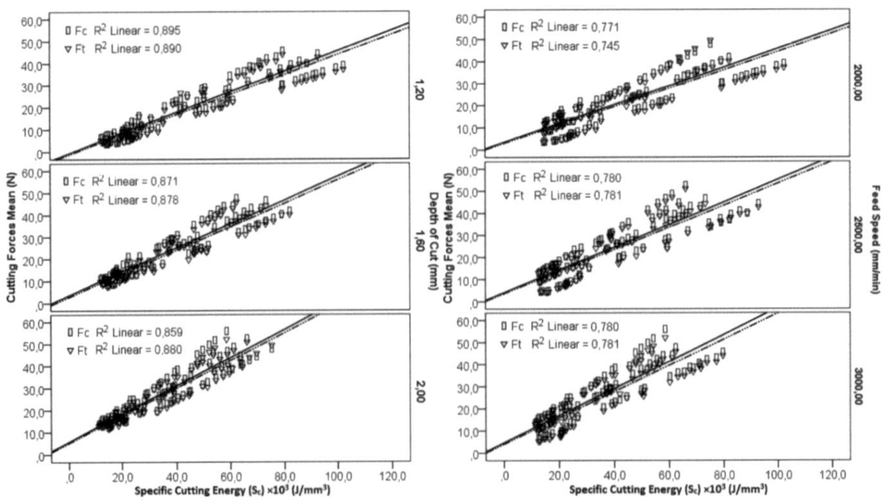

Figure 8.
Relationship between the C_f of natural rocks and S_c according to cutting depth and feed speed.

Proper selection and performance estimation of mill cutting tools are important factors in improving the efficiency of processability and decrease costs in natural rocks. Performance cutting parameters and 3D design of mill cutting tools are cutting tool diameter, d_p, V_a, F_c, and F_t, S_c and S_e. A contribution was made to the literature by proposing new correlation coefficients (R^2) for natural rocks.

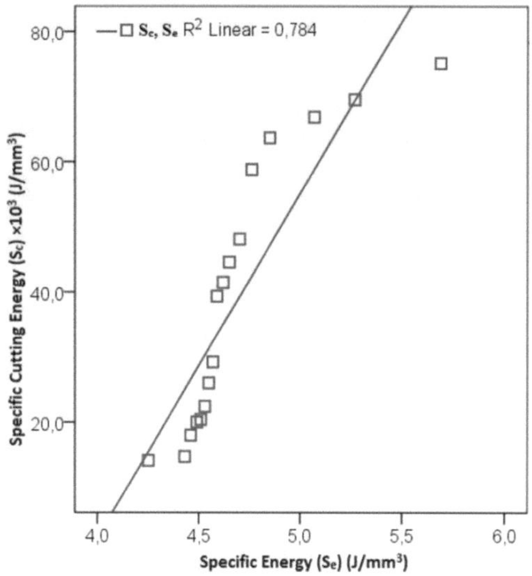

Figure 9.
Relationship between the S_c and S_e of natural rocks.

Independents/ C_f	Dependents/physico-mechanical properties	Custom equation	R^2 Linear Model
F_c	D	y = 3.506 * x + 2625.962	0.931
	P	y = 0.035 * x + 1.595	0.923
	WA	y = −0.031 * x + 1.356	0.887
	KH	y = 1.982 * x + 99.007	0.976
	UCS	y = 1.842 * x + 53.365	0.981
	FS	y = 0.244 * x + 5.129	0.889
	IS	y = 0.547 * x + 14.612	0.968
	AR	y = −0.245 * x + 25.189	0.932
F_t	D	y = 3.588* x + 2627.838	0.930
	P	y = −0.035 * x + 1.575	0.919
	WA	y = −0.031 * x + 1.338	0.883
	KH	y = 2.030 * x + 100.026	0.978
	UCS	y = 1.217 * x + 42.861	0.983
	FS	y = 0.251 * x + 5.237	0.896
	IS	y = 0.560 * x + 14.902	0.968
	AR	y = −0.251 * x + 25.047	0.928

Table 10.
Relationships between F_c and F_t cutting forces natural rock properties.

Independents/S_c and S_e	Dependents/physico-mechanical properties	Custom equation	R^2 linear model
S_c	D	y = 2.017* x + 2630.804	0.929
	P	y = −0.020*x + 1.547	0.922
	WA	y = −0.017 * x + 1.314	0.887
	KH	y = 1.142 * x + 101.696	0.977
	UCS	y = 0.684 * x + 43.858	0.983
	FS	y = 0.140 * x + 5.456	0.892
	IS	y = 0.315 * x + 15.357	0.968
	AR	y = −0.141 * x + 24.845	0.929
S_e	D	y = 115.7 * x + 2166.149	0.858
	P	y = −1.032 * x + 5.610	0.686
	WA	y = −0.883 * x + 4.766	0.616
	KH	y = 63.347 * x + −151.219	0.844
	UCS	y = 37.511 * x + −105.578	0.829
	FS	y = 8.647 * x + −29.663	0.944
	IS	y = 17.500 * x + −54.520	0.837
	AR	y = −7.476 * x + 54.445	0.732

Table 11.
Relationships between S_c and S_e natural rock properties.

4. Conclusion

The rocks in this study were determined according to by taking into account the C_f, S_c and S_e. value in accordance with the statistical analyses. C_f, S_c and S_e values vary depending on the d_p and V_a used in natural rocks. Results of the experimental study are summarized below:

- F_c and F_t values are high at the d_p of 2.0 mm and V_a of 3.000 mm/min.
- S_c and S_e, values are high at the d_p of 1.2 mm and V_a of 2.000 mm/min.
- Due to the increased friction force due to the increase in the amount of natural rock chips to be cut by the cutter edge at the d_p of 2.0 mm, the processability is most difficult.
- The cutting parameters with the highest specific energy volume are considered to be the toughest moments of the CNC machine.
- The values of S_c and S_e are reduced and the efficiency increases with an increase in the d_p which results from the increase in the cutting edge of the

chip volume in natural rocks. S_c and S_e are significantly increased at the d_p of 1.2 mm.

- There is a significant relationship between C_f, S_c and S_e depending on the d_p and V_a, the correlation coefficient (R^2) values of which are 0.859 to 0.895, and 0.745 to 0.781 respectively.
- A significant relationship between the S_c and S_e was identified as R^2 (0.784).
- There is a significant relationship between the C_f and physicomechanical properties. R^2 ranges from 0.887 to 0.981 in the Fc and from 0.883 to 0.983 in the F_t.
- There is a significant relationship between the S_c, S_e, and physicomechanical properties. R^2 ranges from 0.887 to 0.983 in the S_c and from 0.616 to 0.944 in the S_e.

Acknowledgements

This study was supported (Project number: 13.GÜZSAN.01 and TR33/12/SKMDP/0104). We would like to thank them for their (Afyon Kocatepe University and Zafer Development Agency) support and contributions.

Author details

Gencay Sariisik
Faculty of Engineering, Department of Industrial Engineering, Harran University, Şanlıurfa, Turkey

*Address all correspondence to: gsariisik@gmail.com

References

[1] Sariisik A, Sariisik G, Şentürk A. Applications of glaze and decor on dimensioned andesites used in construction sector. Construction and Building Materials. 2011;**25**(9): 3694-3702

[2] Çoşkun G, Sarıışık G, Sarıışık A. Classification of parameters affecting slip safety of limestones. Cogent Engineering. 2016;**3**(1):1-15. DOI: 10.1080/23311916.2016.1217821

[3] Sarıışık G, Çoşkun G, Sarıışık A. Slip safety risk analysis of surface properties using the coefficients of friction of rocks. International Journal of Occupational Safety and Ergonomics. 2019;**25**(3):1-15

[4] Senturk A, Gunduz L, Tosun YL, Sariisik A. Marble Technology. Turkey: Tugra Press; 1996. p. 242. (in Turkish)

[5] Kulaksız S. Natural Stone (Marble) Mining and Processing Technologies. Turkey: TMMOB Chamber of Mining Engineers; 2007. p. 634. (in Turkish)

[6] Sariisik A, Sariisik G. Environmental interaction properties of marble used in the restoration of historical monuments (Dalyan-Kaunos). Ekoloji. 2011;**79**: 12-20

[7] Sarıışık G. Oyman E. Marble waste assessment of stone processıng machınery computer controlled (CNC) use wıth new product development I. Mermer Artıklarının Değerlendirilmesi ve Çevresel Etkilerinin Azaltılması Sempozyumu (16–17 October 2009) Diyarbakır. Turkey; 2009

[8] Sarıışık G. Özkan E. Computer-Controlled Machines (CNC) with the Processing of Natural Stone Wastes and Modeling. 7. Ulusal Kırmataş Sempozyumu (3–4 March 2015). İstanbul. Turkey; 2015

[9] Romoli L. Cutting force monitoring of chain saw machines at the variation of the rake angle. International Journal of Rock Mechanics and Mining Sciences. 2018;**101**:33-40

[10] Copur H, Bilgin N, Balci C, Tumac D, Avunduk E. Effects of different cutting patterns and experimental conditions on the performance of a conical drag tool. Rock Mechanics and Rock Engineering. 2017;**50**(6): 1585-1609

[11] Almasi SN, Bagherpour R, Mikaeil R, Ozcelik Y. Analysis of bead wear in diamond wire sawing considering the rock properties and production rate. Bulletin of Engineering Geology and the Environment. 2017;**76**(4):1593-1607

[12] Yarahmadi R, Bagherpour R, Khademian A, Sousa LM, Almasi SN, Esfahani MM. Determining the optimum cutting direction in granite quarries through experimental studies: A case study of a granite quarry. Bulletin of Engineering Geology and the Environment. 2019;**78**(1):459-467

[13] Özkan E, Sarıışık G, Ceylan S. Application and productivity analysis of new channel opening method in natural stone quarries with diamond wire cutting machine. Arabian Journal of Geosciences. 2015;**8**(2):1089-1098

[14] Zichella L, Bellopede R, Marini P, Tori A, Stocco A. Diamond wire cutting: A methodology to evaluate stone workability. Materials and Manufacturing Processes. 2017;**32**(9):1034-1040

[15] Faria RF, Muniz EP, Oliveira LGS, Proveti JRC, da Silva Porto PS. Using on site data to study efficiency in industrial granite cutting. Journal of Cleaner Production. 2017;**166**:1113-1121

[16] Tumac D, Avunduk E, Copur H, Balci C. Investigation of the effect of

textural properties towards predicting sawing performance of diamond wire machines. Rock Mechanics and Rock Engineering: From the Past to the Future. 2016:211-215

[17] Hasanpour R, Ozcelik Y, Yilmazkaya E, Sohrabian B. DEM modeling of a monowire cutting system. Arabian Journal of Geosciences. 2016; **9**(20):739

[18] Turchetta S, Carrino L, Polini W. CVD diamond insert in stone cutting. Diamond and Related Materials. 2005; **14**(3):641-645. DOI: 10.1016/j.diamond.2004.10.031

[19] Bayram F. Prediction of sawing performance based on index properties of rocks. Arabian Journal of Geosciences. 2013;**6**(11):4357-4362. DOI: 10.1007/s12517-012-0668-5

[20] Ersoy A, Atıcı U. Specific energy prediction for circular diamond saw in cutting different types of rocks using multivariable linear regression analysis Journal of Mining Science. 2005;**41**(3): 240-260. DOI: 10.1007/s10913-005-0089-x

[21] Yilmazkaya E, Ozcelik Y. The effects of operational parameters on a mono-wire cutting system: Efficiency in marble processing. Rock Mechanics and Rock Engineering. 2016;**49**(2):523-539

[22] Almasi SN, Bagherpour R, Mikaeil R, Ozcelik Y. Developing a new rock classification based on the abrasiveness. hardness. and toughness of rocks and PA for the prediction of hard dimension stone sawability in quarrying. Geosystem Engineering. 2017;**20**(6):295-310

[23] Yurdakul M, Akdas H. Prediction of specific cutting energy for large diameter circular saws during natural stone cutting. International Journal of Rock Mechanics and Mining Sciences. 2012;**53**:38-44. DOI: 10.1016/j.ijrmms.2012.03.008

[24] Jerro HD, Pamg SS, Yang C, Mirshams RA. Kinematics analysis of the chipping process using the circular diamond saw blade. Transactions of ASME: Journal of Manufacturing Science and Engineering. 1999;**121**(2): 257-264. DOI: 10.1115/1.2831214

[25] Turchetta S. Cutting force in stone machining by diamond disk. Advances in Materials Science and Engineering. 2010;**63**:1-6. DOI: 10.1155/2010/631437

[26] Turchetta S, Polini W. Cutting force in stone lapping. International Journal of Advanced Manufacturing Technology. 2011;**57**(5–8):533-539

[27] Gelfusa G, Turchetta S. Cutting force and tool wear of single diamond-coated bead. International Journal of Advanced Manufacturing Technology. 2014;**72**(5–8):1063-1072

[28] Asche J, Tönshoff HK, Friemuth T. Cutting principles. wear and applications of diamond tools in the stone and civil engineering industry. In: Proceedings of Diamond Tools Conference. 1999. pp. 151-157

[29] Turchetta S, Sorrentino L, Bellini C. A method to optimize the diamond wire cutting process. Diamond and Related Materials. 2017;**71**:90-97. DOI: 10.1016/j.diamond.2016.11.016

[30] Gelfusa G, Turchetta S. Cutting force and tool wear of single diamond-coated bead. International Journal of Advanced Manufacturing Technology. 2014;**72**(5–8):1063-1072

[31] Turchetta S, Polini W, Gelfusa G, Venafro E. A new sawing machine by diamond wire. The International Journal of Advanced Manufacturing Technology. 2014;**70**(1–4):73-78. DOI: 10.1007/s00170-013-5247-8

[32] Tönshoff HK, Hillmann-Apmann H, Asche J. Diamond tools in stone and civil engineering industry: Cutting

principles. wear and applications. Diamond and Related Materials. 2002; **11**(3):736-741. DOI: 10.1016/S0925-9635(01)00561-1

[33] Bulut B, Tazegul O, Baydogan M, Kayali ES. The comparison of the sintering methods for diamond cutting tools. Journal of Achievements in Materials and Manufacturing Engineering. 2016;**76**(1):30-35

[34] Akdas H, Yurdakul M. Analysis of the industrial cutting process of natural building stones: Evaluation of electric power consumption. Journal of Testing and Evaluation. 2014;**42**(4):931-941. DOI: 10.1520/JTE20130146

[35] Yurdakul M, Gopalakrishnan K, Akdas H. Prediction of specific cutting energy in natural stone cutting processes using the neuro-fuzzy methodology. International Journal of Rock Mechanics and Mining Sciences. 2014;**67**:127-135

[36] Polini W, Turchetta S. Force and specific energy in stone cutting by diamond mill. International Journal of Machine Tools and Manufacture. 2004; **44**(11):1189-1196. DOI: 10.1016/j.ijmachtools.2004.04.001

[37] Turchetta S. Cutting force and diamond tool wear in stone machining. International Journal of Advanced Ma nufacturing Technology. 2012;**61**(5–8): 441-448. DOI: 10.1007/s00170-011-3717-4

[38] Sarıışık G, Özkan E. Mermerlerin CNC Makinesi ile İşlenmesinde Kesme Kuvvetleri ve Spesifik Kesme Enerjisinin İstatistiksel Analizi. Journal of Science and Engineering. 2017; **19**(55):178-193. DOI: 10.21205/deufmd.2017195514

[39] Sarıışık G, Özkan E. Effects of natural rock properties on cutting forces. specific energy and specific cutting energy by four-axis machine. Arabian Journal of Geosciences. 2018; **11**(5):84. DOI: 10.1007/s12517-018-3424-7

[40] Sarıışık G, Öğütlü AS. İstatistiksel K-Ortalamalar Kümeleme Analizi ile Doğal Taşların Yeni İşlenebilirlik İndeksi. Harran Üniversitesi Mühendislik Dergisi. 2018;**3**(3):156-165

[41] Turkish Standards Institute (TSI). Natural Stone Test Methods - Determination of Real Density and Apparent Density. and of Total and Open Porosity. Ankara: TSI; 2010. Standard No. TS EN 1936: 2010. Turkish

[42] Turkish Standards Institute (TSI). Natural Stone Test Methods— Determination of Water Absorption at Atmospheric Pressure. Ankara: TSI; 2014. Standard No. TS EN 13755: 2014. Turkish

[43] Turkish Standards Institute (TSI). Natural Stone Test Methods - Determination of Knoop Hardness. Ankara: TSI; 2004. Standard No. TS EN 14205: 2004. Turkish

[44] Turkish Standards Institute (TSI). Natural Stone Test Methods- Determination of Compressive Strength. Ankara: TSI; 2007. Standard No. TS EN 1926: 2007. Turkish

[45] Turkish Standards Institute (TSI). Natural Stone Test Methods- Determination of Flexural Strength under Constant Moment. Ankara: TSI; 2014. Standard No. TS EN 13161: 2014. Turkish

[46] Turkish Standards Institute (TSI). Natural Stone Test Methods - Determination of Frost Resistance. Ankara: TSI; 2009. Standard No. TS 699: 2009. Turkish

[47] Turkish Standards Institute (TSI). Natural Stone Test Methods Determination of Abrasion Resistance. Ankara: TSI; 2013 Standard No. TS EN 1341 Appendix-C: 2013. Turkish

Chapter 3

Advanced Mapping of the Seafloor Using Sea Vehicle Mounted Sounding Technologies

Tongwei Zhang, Baohua Liu and Xiaodong Liu

Abstract

A large proportion of the Earth's surface is the deep sea. Numerous fields require access to seafloor topography and geomorphology. With the emergence of different types of underwater vehicles, especially the commercialization of near-seafloor micro-topographical mapping sonars, near-seafloor micro-topographical detection in the deep sea is possible. Near-seafloor micro-topographical exploration allows accurate detection of the seafloor using multibeam echosounder, side-scan sonar, and bathymetric side-scan sonar carried on-board various vehicles, including deep-tow, autonomous underwater vehicles, remotely operated vehicles, and human-occupied vehicles. Near-seafloor micro-topographical detection can obtain more accurate micro-topography and micro-geomorphology of the seafloor compared to full sea depth topographical detection. In this chapter, the basic principles of three types of near-seafloor micro-topographical mapping sonars are analyzed. Then, four types of underwater vehicles that are suitable for near-seafloor micro-topographical mapping are briefly discussed. Factors affecting mapping and detection results are presented using the Jiaolong human-occupied vehicle and its bathymetric side-scan sonar as an example. Next, the entire data processing and mapping methods are described. Finally, two typical detection results obtained by the Jiaolong bathymetric side-scan sonar in deep-sea are given.

Keywords: topography, geomorphology, deep sea, underwater vehicle, mapping sonar

1. Introduction

The ocean accounts for about 71% of the Earth's surface area and is the largest potential resource base on the Earth that has not been fully recognized and utilized by humans. There are extremely rich biological and mineral resources in the ocean. The deep sea is the lowest layer in the ocean, existing below the thermocline, at a depth of 1800 m or more [1]. Deep sea areas with a depth of more than 2000 m account for 84% of the ocean area. Therefore, the surface of the earth is mostly deep sea. The hypsometric profile of the ocean is shown in **Figure 1**.

Scientific research, resource development, engineering construction, and military activities around the ocean usually require accurate acquisition of seafloor topographical information in the area of interest as the basis for data and support. Therefore, understanding ocean topographical information, mapping ocean

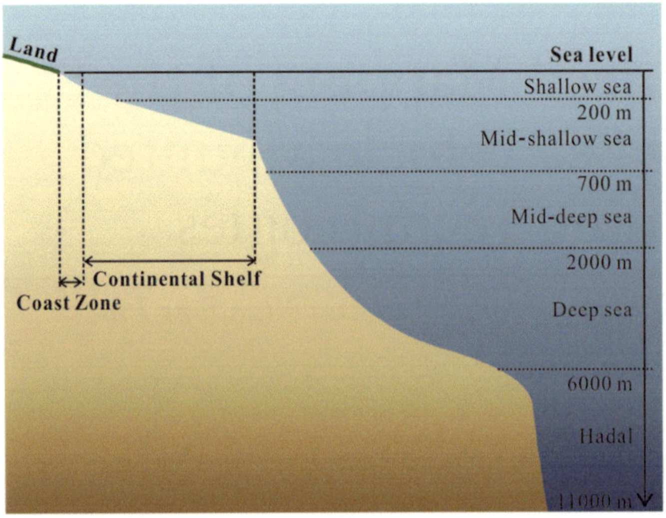

Figure 1.
Hypsometric profile of the ocean.

topographical and geomorphological information effectively, and how to obtain ocean topographical information map have become important issues in the current development of marine resources and marine space utilization [2]. In particular, marine topographical information is of immense value in marine space utilization.

Topography involves the recording of terrain, the three-dimensional quality of the surface, and the identification of specific landforms. It is often considered to include the graphic representation of the landform on a map by a variety of techniques, including contour lines, hypsometric tints, and relief shading [3]. Geomorphology is the branch of science that studies the characteristics and configuration and evolution of rocks and landforms [4]. In the ocean, seafloor topography is measured by multibeam echosounder (MBE), and seafloor geomorphology is measured by side-scan sonar (SSS).

Deep sea topographical exploration mainly includes full sea depth topographical detection and near-seafloor micro-topographical detection. The advantage of full sea depth topographical is large spatial range and rapid data acquisition, and the disadvantage is limited accuracy. In contrast, near-seafloor micro-topographical exploration provides accurate detection of the seafloor using MBE, SSS, and bathymetric side-scan sonar (BSSS) carried on-board various underwater vehicles, including deep tow (DT) [5], autonomous underwater vehicle (AUV) [6], remotely operated vehicle (ROV) [7], and human occupied vehicle (HOV) [8–10]. It can obtain more accurate micro-topography and micro-geomorphology of the seafloor compared to full sea depth topographical detection.

In this chapter, the basic principles of three types of near-seafloor micro-topographical mapping sonars are analyzed. Then, four types of underwater vehicles suitable for near-seafloor micro-topographical mapping are briefly discussed. Next, factors affecting mapping and detection results are presented using the Jiaolong HOV and its BSSS as an example. Finally, the entire data processing and mapping methods are presented.

2. Near-seafloor micro-topographical mapping sonars

Three types of seafloor mapping sonars, which can be mounted of deep sea vehicles, are the multibeam echosounder (MBE), and the side-scan sonar (SSS) and the bathymetric side-scan sonar (BSSS).

2.1 Multibeam echosounder

An MBE works by transmitting a wide-sector-covered sound wave to the seafloor using a transmitting transducer array, and the narrow-beam receives the sound wave using a receiving transducer array. The footprints of the seafloor topography are formed by the orthogonality of the transmission and reception sectors, and these footprints are properly processed. A ping can indicate the water depth values of hundreds or even more seafloor measured points in the vertical plane perpendicular to the heading. Therefore, it is possible to accurately and quickly measure the size, shape, and height variation of underwater targets within a certain width of the route, and to reliably depict the three-dimensional features of the seafloor topography. The basic principle of an MBE is shown in **Figure 2**.

The beamforming method of MBE can be divided into two types: beam steering method (measuring the round-trip time of the reflected signal at a specific angle) and coherent method (measuring the angle of the reflected echo signal at a specific time). There are two main variables to be measured in an MBE, namely the slant distance or the distance from the acoustic transducer to each point on the seafloor and the angle from the transducer to the bottom of the ocean. All MBEs use one or both beamforming methods to determine these variables. At present, MBE manufacturers using beam steering method include Reson, Kongsberg, ATLAS, L3, and R2Sonic, whereas manufacturers using coherent method include Teledyne Benthos and Geoacoustics.

For large-area exploration of seafloor topography, shipborne deep-water MBE can be used to obtain relatively accurate seafloor topographical data. The frequency of deep-water MBE is generally approximately 12 kHz. A typical beam width is 1°× 1°, and the corresponding beam footprint is 1.75% water depth. For example, when the water depth is 5000 m, the beam footprint is approximately 87.5 m. It can be observed that the shipborne deep-water MBE cannot obtain high-precision seafloor topographical data.

Underwater vehicles, such as DT, AUV, ROV, and HOV, can carry more high-frequency MBE to near-seafloor to achieve accurate topographical detection. The corresponding MBE is designed with a special pressure-resistant design with a pressure depth of up to 6000 m or even deeper. At present, some commercial

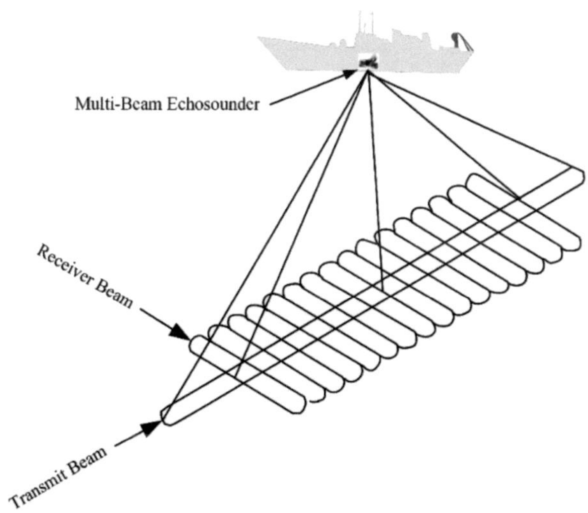

Figure 2.
Basic principle of MBE.

MBEs used for deep sea underwater vehicles are Kongsberg EM2040, Reson SeaBat 7125/T20-S, and R2Sonic 2022/4/6. They are widely used on different underwater vehicles around the world. They are widely used on different underwater vehicles around the world. Comparison of typical high-resolution MBEs for deep sea underwater vehicles is shown in **Table 1**.

2.2 Side-scan sonar

SSS generates an acoustic image by emitting an acoustic signal and receiving an echo signal reflected by the seafloor to reveal sea bottom conditions, including the position, current status, height, and shape of the target. SSS has advantages of intuitive image, high resolution, and large coverage compared to other seafloor detection technologies.

SSS can be divided into two types according to the installation position of the acoustic transducer array: shipborne type and towed type. A shipborne acoustic transducer array is mounted on both sides of the ship hull. This type of SSS operates at a generally lower frequency (below 10 kHz) and has a wider swath. On the other hand, a towed acoustic transducer array is installed in the tow body, only a few tens of meters away from the seafloor, and the speed is low. The obtained side-scan image quality is higher, and even a pipeline of 10 cm and a small volume of oil drum can be distinguished. Recently, the speeds of some deep tow type SSS systems have increased, and high-resolution side-scan images can still be obtained at 10 kn.

SSS technology has two development directions: one direction is to develop BSSS technology that can obtain the topography of the seafloor while obtaining the seafloor geomorphology and the other direction is the development of synthetic aperture sonar technology with lateral resolution theoretically equal to half the physical length of the sonar array and does not increase with increasing distance.

At present, commercial SSSs for deep-sea underwater vehicles commonly used are Klein 3000, EdgeTech 2200-M, and Kongsberg dual-frequency sonar. A comparison of typical SSSs for deep sea underwater vehicles is shown in **Table 2**.

2.3 Bathymetric side-scan sonar

In order to incorporate the advantage of MBE and SSS, the Institute of Acoustics of the Chinese Academy of Sciences (IOACAS) [11] developed BSSS. BSSS can detect seafloor geomorphology and topography simultaneous. The arrival angle

	Kongsberg, EM2040		Reson, SeaBat 7125		R2Sonic, Sonic 2024	
Frequency	200 kHz	400 kHz	200 kHz	400 kHz	200 kHz	450 kHz
Transmit beamwidth	0.7°	0.4°	2.0°	1.0°	1.0°	0.45°
Receive beamwidth	1.5°	0.7°	1.0°	0.5°	2.0°	0.9°
Depth	635 m	315 m	450 m	175 m	400 m	—
Coverage	200°	200°	165°	165°	160°	160°
Number of beams	400	400	256	512	1024	1024
Ping rate	60 Hz		50 Hz	50 Hz	60 Hz	60 Hz
Range resolution	14.2 mm	10.5 mm	6 mm	6 mm	—	10.2 mm
System depth rating	6000 m		6000 m		6000 m	

Table 1.
Comparison of typical high-resolution MBEs for deep sea underwater vehicles.

and the water depth of seafloor echoes can be measured by receiving arrays of BSSS. The advantage of BSSS is high resolution, small array, and low power. **Figure 3** shows the basic principles of BSSS.

The BSSS system's small size, lightweight, and low power consumption make it especially suitable for installation on DT, AUV, ROV, and HOV.

	Klein, 3000	EdgeTech, 2200-M	Kongsberg, dual-frequency sonar
Frequency	100/500 kHz	75/410 kHz 120/410 kHz 75/120 kHz 300/600 kHz	114/410 kHz
Horizontal beams	0.7° (100 kHz) 0.2° (500 kHz)	0.6° (120 kHz) 0.3° (410 kHz)	1.0° (114 kHz) 0.3° (410 kHz)
Beam tilt	5/10/15/20/25°	20°	10° ± 1°
Range	600 m (100 kHz) 150 m (500 kHz)	500 m (120 kHz) 150 m (410 kHz)	600 m (114 kHz) 150 m (410 kHz)
Depth rating	3000 m	6000 m	2000 m

Table 2.
Comparison of typical SSSs for deep sea underwater vehicles.

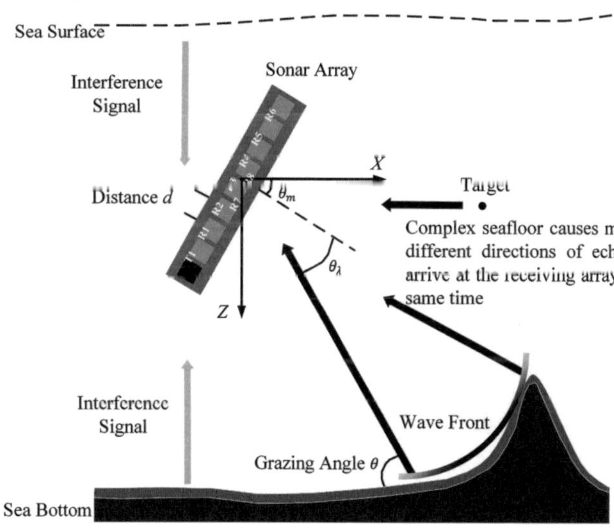

Figure 3.
Basic principle of BSSS.

	IOACAS HRBSSS	Teledyne, Benthos C3D	Kongsberg, GeoSwath Plus
Frequency	150 kHz	200 kHz	125–500 kHz
Bathymetry coverage	2 × 300 m	10–12 × depth	100 m
Side-scan coverage	2 × 400 m	2 × 300 m	200 m
Speed	2.5 kn	3–5 kn	3 kn
Depth rating	7000 m	6000 m	4000 m

Table 3.
Comparison of typical BSSSs for deep sea underwater vehicles.

At present, commercial BSSS for deep-sea underwater vehicles mainly includes IOACAS HRBSSS, Teledyne Benthos C3D, and Kongsberg GeoSwath Plus. The high-resolution BSSS developed by IOACAS can simultaneously obtain high-resolution seafloor topography and geomorphology, and is suitable for use in complex conditions in the deep sea. HRBSSS has been successfully applied to various deep-sea underwater vehicles such as Jiaolong HOV, DTA-6000 acoustic DT, and Qianlong I/II AUV. A comparison of typical BSSSs for deep sea underwater vehicles is shown in **Table 3**.

3. Deep sea vehicles

Four types of vehicles, on which seafloor mapping sonars can be mounted of deep sea surveys are the deep tow, the automated underwater vehicle (AUV), the remotely operated vehicle (ROV), and the human occupied vehicle (HOV).

3.1 Deep tow

A deep tow (DT) system is a large-scale marine equipment used for surveying the characteristics of the seafloor in the deep sea. It is generally placed near-seafloor using a towing cable to investigate high-precision topography, superficial geological structures, flow fields, and other physical and chemical parameters. A DT system consists mainly of a deck control unit, a winch, cables, a depressor, and a tow body, as shown in **Figure 4**. The A-shaped frame on the mothership is used to drop the tow body into the sea. The towing speed is less than 5 kn. In general, the distance between the tow body and the seafloor is controlled to approximately 50 m, and the mothership moves to pull the DT for topographical and hydrological data collection. The collected data are transmitted to the deck control unit of the shipboard laboratory for processing using a winch and armored photoelectric composite cable. The mothership supplies power to the tow body using the armored photoelectric composite cable. Therefore, the DT system can be used for long-term continuous operation and is widely used in deep-sea survey, target search, and other fields.

The tow body can be used as a carrier to carry topographical or hydrological survey equipment as needed. Micro-topographical mapping sonars (such as MBE, SSS, BSSS, and sub-bottom profiler) installed on the DT can detect seafloor topography near the seafloor and obtain more accurate topography and geomorphology of the seafloor.

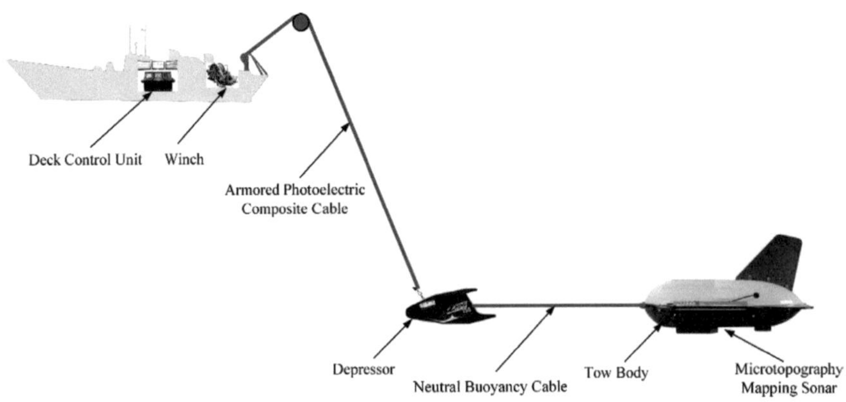

Figure 4.
Deep tow system

3.2 Autonomous underwater vehicle

AUVs (**Figure 5**) are autonomous underwater vehicle that can move in the ocean without real-time control by human being. The AUV itself carries energy and detection equipment and can achieve autonomous navigation and detection operations, less manual intervention, and low mother ship requirements. It is very suitable for large-area underwater exploration and data acquisition operations. The AUV is equipped with relevant sensors for different detection purposes, including acoustic, chemical, hydrological, and geophysical sensors. With the maturity of AUV technology, the AUV is currently equipped with micro-topographical mapping sonars, such as MBE, SSS, and BSSS to near-seafloor to measure underwater terrain as a new option.

3.3 Remotely operated vehicle

A remotely operated vehicle (ROV, **Figure 6**) is an unoccupied underwater robot that is connected to a mother ship by a series of cables. A load-carrying umbilical cable is used along with a tether management system (TMS) when the ROV works in rough conditions or in deeper water.

Before the ROV dives, it requires a relatively accurate understanding of the topography of the working area. Usually, the topographical map of the working area is obtained by a shipborne deep-water MBE. However, for deep sea areas, the accuracy of shipborne deep-water MBE is limited. To ensure safety of the ROV when the topography of the seafloor is complex, it is necessary to use micro-topographical

Figure 5.
Autonomous underwater vehicle.

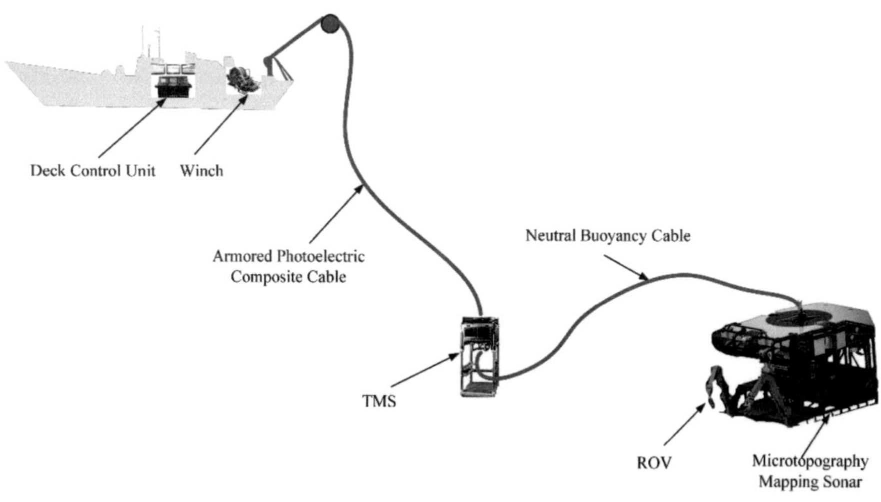

Figure 6.
Remotely operated vehicle.

mapping sonars mounted on the ROV to obtain fine topographical map of the working area. However, unlike DT and AUV, ROV is not suitable for large-area mapping surveys.

3.4 Human occupied vehicle

HOV is similar to a small submarine; however, a submarine is not submersible. Equipment on an HOV typically includes a manipulator, a camera system, and a special lighting system. HOV is a more versatile underwater vehicle than a submarine, which is reflected in several applications. An HOV is designed to dive to greater depths, just like a submarine. Due to enormous pressures in the deep sea, it requires a special pressure-resistant design that carries no more than two or three people and limited food, water, and oxygen.

According to the database of the Manned Underwater Vehicles Professional Committee of the Marine Technology Society (MTS) of the United States, there are currently 16 HOVs worldwide with a depth capacity of over 1000 m [12]. These include the Alvin (4500 m; United States), Nautile (6000 m; France), Mir I and Mir II (6000 m; Russia), Shinkai 6500 (6500 m; Japan), and Jiaolong (7000 m; China).

Jiaolong is equipped with a BSSS system to obtain accurate mapping of seafloor topography and geomorphology, similar to other great-depth HOVs, as shown in **Figure 7**. The BSSS system consists of two parts: one part is installed in the manned cabin (i.e., the master controller unit), whereas the other part is installed outside the manned cabin (i.e., the electronic cabin, port-side transducer array, starboard-side transducer array, and subsidiary sensors).

The major axis of the BSSS transducer array must be parallel to the major axis of the HOV. The transducer surface normal must have an angle of 30° above the horizontal plane. In addition, the transducer array is installed between HOV stations 4 and 5 as deformation of the mounting bracket must be minimized during lifting to avoid damaging the transducer array. In this cylindrical part of the HOV, the transducer array has a better line-type after installation; the mounting bracket is independent of the load-bearing frame, thereby reducing the impact of frame deformation on the transducers.

BSSS is mainly used to obtain data on micro-topography and micro-geomorphology: ultra-short baseline (USBL) and long baseline (LBL) provide navigation and positioning data, which are essential for topographical and geomorphological mapping; underwater acoustic communication devices transmit positioning data obtained by USBL on the supporting mothership at the surface to HOV, allowing the

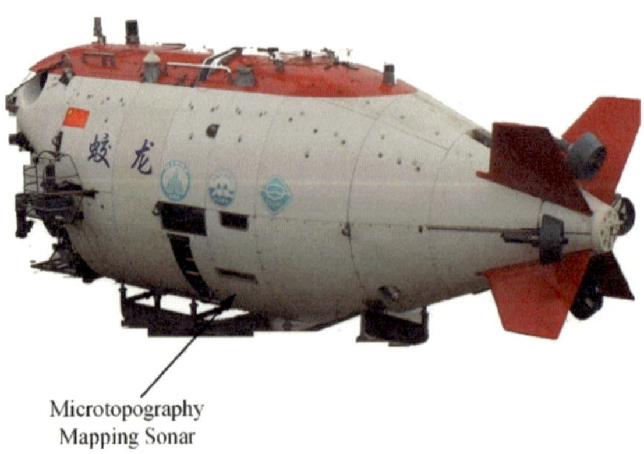

Figure 7.
Human-occupied vehicle Jiaolong.

determination of the initial position of the integrated navigation system; the HOV speed is measured by Doppler velocity log (DVL). The attitude and heading are measured by optical fiber compass. The sound speed is provided by the conductivity-temperature-depth (CTD). The depth of HOV is measured by a high-accuracy depth sensor, which combined with topographical data measured by BSSS, the absolute depth of the seafloor can be obtained.

4. Factors affecting mapping and detection results

In general, factors affecting data quality when using small underwater vehicles (e.g., DT and AUV) to carry near-seafloor micro-topographical mapping sonars fall into five categories: horizontal positioning accuracy, vertical positioning accuracy, depth accuracy, sensor time uniformity (time requirements of the sensor are uniform, the attitude sensor time is accurate to 50 ms, the transmission time is accurate to 50 ms, and the other sensors are accurate to 1 s), and sensor location uniformity (it requires precise knowledge of the coordinates of each sensor relative to the origin of the coordinates origin and installation errors). Large-scale underwater vehicles (e.g., ROV and HOV) not only have the above five features but also have other additional characteristics, including poor stability in attitude control, acoustic transducer array port and starboard installations, and wide spacing. Hence, detailed discussion on factors affecting mapping and detection results is presented using Jiaolong HOV and its BSSS as an example. By processing the BSSS detection data collected by the Jiaolong HOV, we found that factors that mostly affect mapping quality are the HOV attitude, port and starboard positions of BSSS transducer array [7].

4.1 Effect of HOV attitude

Because the streamline of HOV and manual control, the attitude control stability of HOV is relatively poor; therefore, HOV attitude is a main factor that effect BSSS detection and mapping. **Figure 8** shows the seafloor micro-topography around the peak of a cold spring area. **Figure 8a** uses raw data; **Figure 8b** uses data in which filtering and smoothing have been applied to the roll angle.

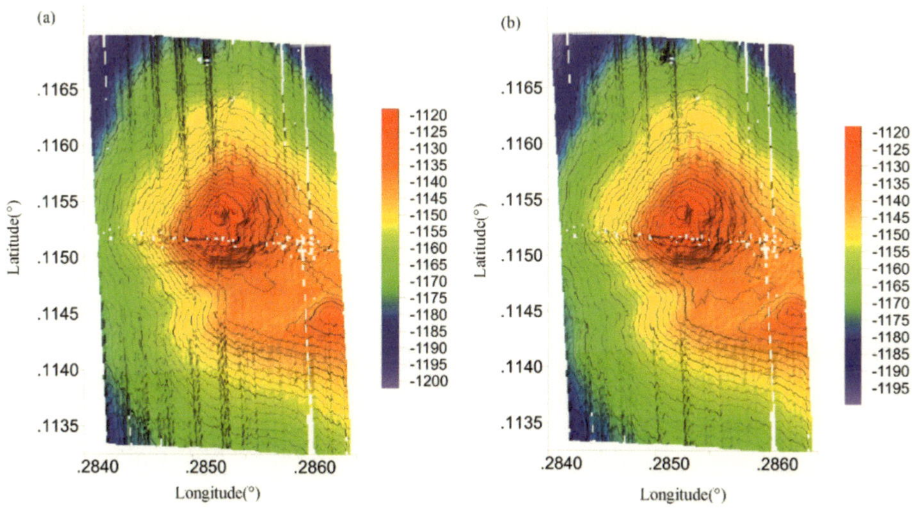

Figure 8.
Seafloor micro-topography around the peak of a cold spring area: (a) map using raw data and (b) map using data in which filtering and smoothing have been applied to the roll angle.

Figure 9 shows the temporal change curves recorded by BSSS for (a) roll angle, (b) pitch angle, and (c) heading angle. It can be found that there is a strong correlation between map distortion in **Figure 8a** and roll angle in **Figure 9a**.

4.2 Effect of port and starboard positions

Seafloor micro-topographical detection involves multiple sensors, including LBL, USBL, DVL, fiber optic compass (FOC), and BSSS arrays. The distance of the two

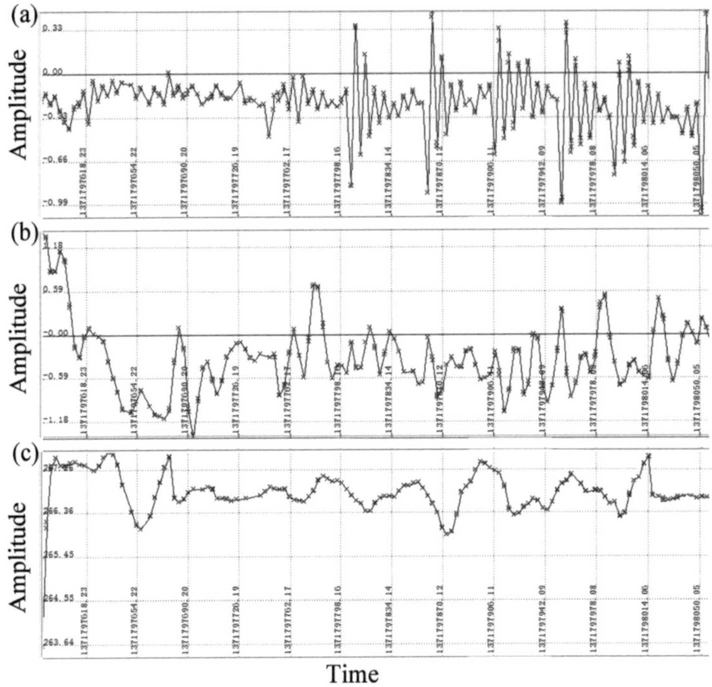

Figure 9.
Temporal change curves recorded by BSSS for (a) roll angle, (b) pitch angle, and (c) heading angle.

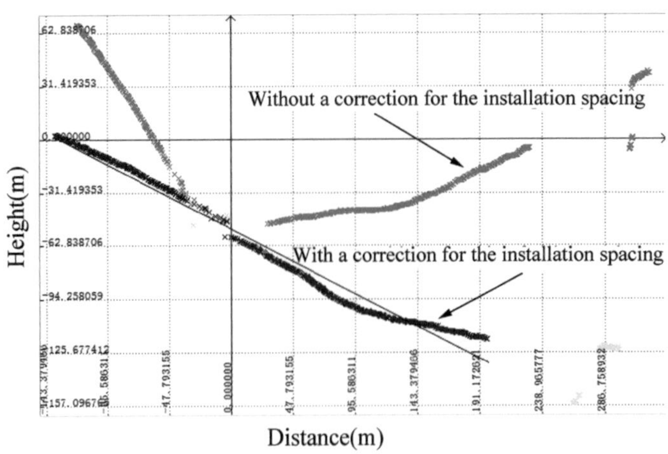

Figure 10.
Depth-sounding results for a single ping measurement along sloped terrain in a cold spring area. The gray dotted line indicates data without correction for the installation spacing between the port and starboard transducer arrays, whereas the black dotted line indicates data with correction for the installation spacing.

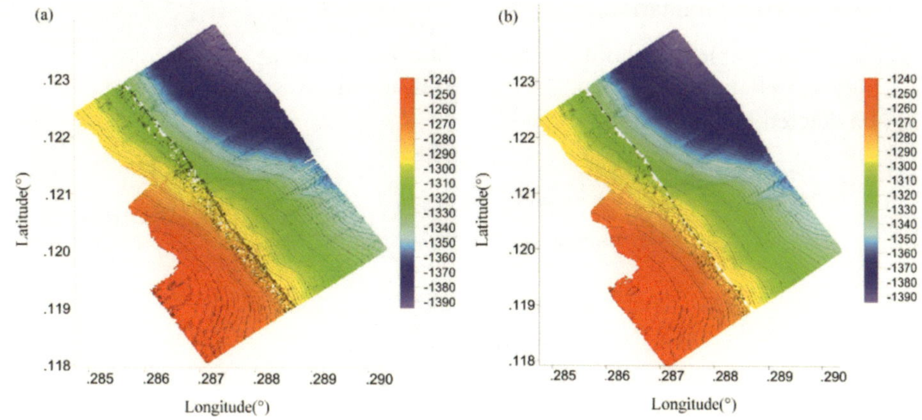

Figure 11.
Seafloor micro-topographical maps of sloped terrain in a cold spring area: (a) without compensation of the array spacing (b) with compensation of the array spacing [10] (permissions obtained to reprint).

BSSS array is about 2.46 m. As BSSS detect the bottom by the two BSSS array separately, so when the bottom has a large slope, the bottom detection results are different.

Figure 10 shows depth results from a single ping measurement of sloped terrain in a cold spring area, with and without correction for the installation spacing. The detection results of two BSSS arrays are different. If there is no compensation, the quality of seafloor topographical map will be poor.

Figure 11a shows that when the spacing of two BSSS arrays is not taken into account, the quality of the mapping result is poor. After compensation of the spacing, the transition of the mapping result is relatively smooth, and the mapping quality is improved, as shown in **Figure 11b**.

5. Data processing and mapping method

The raw data, including BSSS mapping data, HOV attitude data, acoustic positioning data, sound velocity data, are processed in two steps (pre-processing and post-processing) to obtain topographical and geomorphology map. Data pre-processing mostly consists of five steps: navigation data processing, rewriting navigation data, coarse error elimination, angular deviation correction, and port/starboard position correction. Data post-processing mostly consists of 12 steps, namely installation angle correction, sound velocity correction, attitude data filtering, bathymetric data filtering, fine error elimination, field correction, side-scan data filtering, bottom tracking, angle variation gain, equalization gain, data export, and mapping. **Figure 12** shows the flow diagram of data processing and mapping.

5.1 Data pre-processing

1. Navigation data processing

   Post-process navigation and acoustic positioning data, such as FOC, DVL, USBL, and LBL, to obtain high-precision navigation data.

2. Rewriting navigation data

   Rewrite high-precision navigation data to the BSSS raw data file to replace original navigation data.

3. Coarse error elimination

   Set a reasonable threshold for the side-scan data energy. If the energy of side-scan data is lower than the threshold, it indicates a bad ping and should be deleted.

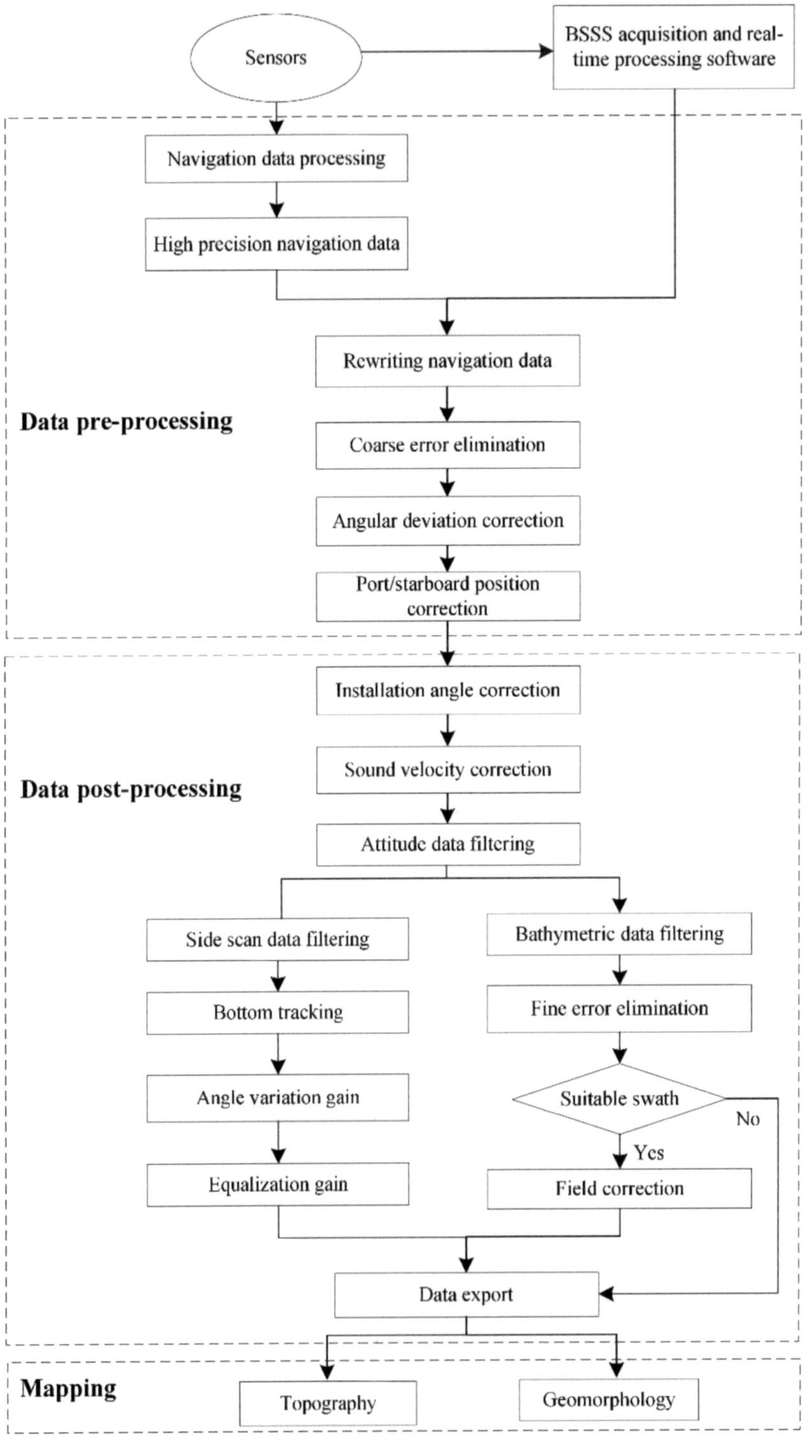

Figure 12.
Flow diagram of data processing and mapping.

4. Angular deviation correction

   Analyze existing BSSS data and select data corresponding to the seafloor with a constant slope. Let the slope pass the point just below the sonar array to find the optimal slope of the seafloor; use this slope to obtain the angle correction curve of bathymetric data. Next, use median filter to filter the obtained angle correction curve, filter out high-frequency components, and retain only low-frequency components to reflect the larger change trend. Finally, use the angle correction curve to perform angle correction on each ping data.

5. Port/starboard position correction

   Read ping data in the BSSS data file, obtain the delay and angle of each ping, and convert to slant range and angle. Calculate the horizontal distance and the vertical height of each ping according to the triangular geometric relationship (taking the installation angle of the sonar array into consideration). Add the port/starboard position to the horizontal distance while maintaining the vertical height. Recalculate the corresponding slant range and angle according to the triangular geometric relationship and convert to delay and angle.

5.2 Data post-processing

1. Installation angle correction

   Correct the installation angle with 30°, which is measured.

2. Sound velocity correction

   Input the measured sound velocity by CTD.

3. Attitude data filtering

   Perform median filtering and smoothing for roll, pitch, and heading.

4. Bathymetric data filtering

   Filter bathymetric data by setting the height, horizontal coordinates, depth, and slant distance.

5. Fine error elimination

   Trim the pings manually to realize fine error elimination.

6. Field correction

   If there is a suitable swath, select the swath that meets the requirements, carry out the corresponding error test, and make corresponding corrections to suppress deviation of the roll, navigation latency, and pitch.

7. Side-scan data filtering

   Filter side-scan data by setting the height, horizontal coordinates, depth, and slant distance.

8. Bottom tracking

   Side-scan data processing requires bottom tracking to delete the water column and generally selects automatic bottom tracking. If the effect is not appropriate, bottom tracking can be manually set.

9. Angle variation gain

   Angle variation gain is used to compensate for backscattering variations due to system beam pattern irregularities and incident angle changes.

10. Gain equalization

    Gain equalization can be used to address the unevenness of side-scan data due to hardware gain changes or excessive pitch angles during data acquisition.

11. Data export

    Export processed data for mapping.

12. Mapping

    Use Surfer software to perform topographical and geomorphological mapping.

6. Typical detection results

Ten years sea-trial and application have shown that the Jiaolong BSSS can produce high-resolution micro-topographical and micro-geomorphological maps; isobaths can be displayed at 2-m intervals, and many seafloor details are clearly distinguished (**Figure 13**). Furthermore, comparison of micro-topographical and micro-geomorphological maps can yield additional information.

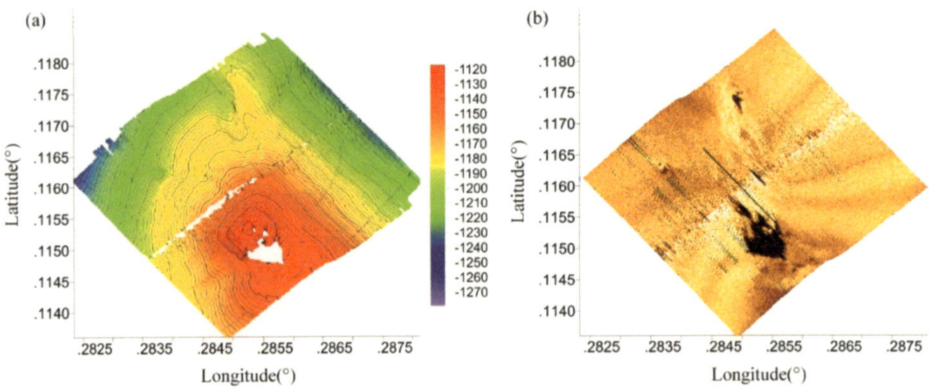

Figure 13.
Seafloor (a) micro-topographical and (b) micro-geomorphological maps obtained by the Jiaolong BSSS on a dive around the peak of a cold spring area [10] (permissions obtained to reprint).

7. Conclusions

Numerous fields require access to seafloor topography and geomorphology. The onboard deep-sea multi-beam sounding system should be used to map the area and obtain a relatively accurate topographic map of the seabed to provide basic terrain data. Although a shipborne deep-sea multibeam sounding system can detect a wide range of deep seabed terrains, its detection accuracy is limited, and it cannot search for small objects on the seafloor.

Near-seafloor micro-topographical mapping is a significant supplement to full sea depth topographical detection. It combines micro-topographical mapping sonars with underwater vehicles and can dive to near-seafloor to achieve accurate topographical detection. DT and AUV are more suitable for large-area micro-topographical mapping survey owing to their good streamline and control stability. However, although the mapping and detection results are affected by other factors, ROV and HOV are still needed for small-site micro-topographical mapping. BSSS and MBE can be mounted on ROV and HOV.

Acknowledgements

This work was supported in part by the Qingdao National Laboratory for Marine Science and Technology under Grant QNLM2016ORP0406, Taishan Scholar Project Funding under Grant TSPD20161007, National Natural Science Foundation of China under Grant 41641049, National Key R&D Program of China under Grant 2017YFC0305700, Shandong Provincial Natural Science Foundation under Grant ZR2015EM005, Shandong Provincial Key R&D Program under Grant 2016GSF115006, and Qingdao Independent Innovation Project under Grant 15-9-1-90-JCH.

Conflict of interest

The authors declare that they have no conflict of interest.

Author details

Tongwei Zhang[1,2,3*], Baohua Liu[1,2] and Xiaodong Liu[4]

1 National Deep Sea Center, Qingdao, China

2 Laboratory for Marine Geology, Qingdao National Laboratory for Marine Science and Technology, Qingdao, China

3 Joint Laboratory for Ocean Observation and Detection, Qingdao National Laboratory for Marine Science and Technology, Qingdao, China

4 Ocean Acoustic Technology Center, Institute of Acoustics, Chinese Academy of Sciences, Beijing, China

*Address all correspondence to: tongwei.zhang@outlook.com

References

[1] Wikipedia contributors. Deep sea. Wikipedia, the free encyclopedia. October 19, 2018, 19:17 UTC. Available from: https://en.wikipedia.org/w/index.php?title=Deep_sea&oldid=864831273. [Accessed: 2018-12-11].

[2] Jin XL. The development of research in marine geophysics and acoustic technology for submarine exploration. Progress in Geophysics. 2007;**22**:1243-1249

[3] Wikipedia contributors. Topography. Wikipedia, the free encyclopedia. December 5, 2018, 14:47 UTC. Available from: https://en.wikipedia.org/w/index.php?title=Topography&oldid=872158696 [Accessed: 2018-12-11]

[4] Wikipedia contributors. Geomorphology. Wikipedia, the free encyclopedia. November 2, 2018, 12:52 UTC. Available from: https://en.wikipedia.org/w/index.php?title=Geomorphology&oldid=866921872. [Accessed: 2018-12-11]

[5] Liu XD, Zhao TH, Cao JL, Wang ZB, Hou DY, Zhang DS. General scheme analysis of the lightweight acoustic deep tow system for gas hydrate exploration. Marine Geology Frontiers. 2015;**31**:8-16

[6] Li YM, Li Y, Sheng MW, Sun YS, Zhang GC. Application and development of topographic survey using multibeam echosounder equipped on AUV. Hydrographic Surveying and Charting. 2016;**36**:7-11

[7] Luo BW, Pao YY, Zhou ZJ, Liu J. Explorative study on short range exploration method for deep sea cobalt crust tiny terrain. Metal Mine. 2006;**41**:15-19

[8] Liu BH, Ding ZJ, Shi XP, Yu KB, Li DW, Li BG. Progress of the application and research of manned submersibles used in deep sea scientific investigations. Haiyang Xuebao. 2015;**37**:1-10

[9] Zhu M, Zhang TW, Yang B, Liu YY, Tang JL. Sonar system of Jiaolong human-occupied vehicle. Chinese Science Bulletin. 2014;**59**:3462-3470

[10] Zhang TW, Tang JL, Li ZG, Zhou YB, Wang XX. Use of the Jiaolong manned submersible for accurate mapping of deep sea topography and geomorphology. Science China Earth Sciences. 2018;**61**:1148-1156

[11] Zhu WQ, Liu XD, Zhang DS, Zhang FS, Qin GL, Liao Z, et al. High resolution bathymetric side scan sonar. Ocean Technology. 2006;**24**:29-35

[12] Kohnen W. Review of deep ocean manned submersible activity in 2013. Marine Technology Society Journal. 2013;**47**:56-68

Chapter 4

Identification of Active Faults in Landslide-Prone Regions Using Ground-Penetrating Radar: A Case Study from Bandung, Indonesia

Maman Hermana, Maya Genisa, Luluan A. Lubis and Chow Weng Sum

Abstract

Ground-penetrating radar or georadar is a popular method in engineering and archeology for investigation of objects in shallow subsurface at high resolution. Georadars produce electromagnetic waves which propagate into the subsurface, and its interaction with the dielectric contrast is reflected and recorded in the radargram. It is an environmentally safe and nondestructive method and can be used for monitoring of active faults in the landslide-prone regions. This chapter explains the concept of georadar and its implementation on the detection of the active fault—Lembang fault—located in Bandung, Indonesia. Bandung is a highly populated city with many living around the active fault which poses a high risk of landslides. The Lembang fault was created by tectonic forces during the Pleistocene and has been constantly reactivated by recent volcanic events. It is the largest active fault in West Java, Indonesia, which is located in the midst of a densely populated urban area. A georadar survey using 25 MHz and 50 Hz frequency antenna was conducted to detect the fault in the urban setting. Unix-based seismic software was used to process the electromagnetic signals. The results showed that the georadar method was successful in identifying the active fault with clear imaging of the subsurface structures and basement of the region.

Keywords: fault, georadar, electromagnetic wave, wave propagation, radargram

1. Introduction

Georadar method is commonly used for engineering and archeology since 1980 [1]. The target of this method usually is to image shallow/near surface. The method promises to give a better resolution and accuracy specially to detect the fault system and other subsurface structures in detail. Georadar is based on electromagnetic wave which detects the contrast of dielectric properties of medium. Due to the high frequency, georadar is able to effectively identify the shallow objects with a high resolution.

The instrument of georadar is equipped with transmitter and receiver antennas which has the ability to transmit and receive electromagnetic wave into and from the earth at certain frequency ranges. Data are recorded as time series in two-way time (TWT) manner which after processing can be converted into depth domain by adding the velocity model during processing. **Figure 1** shows the georadar instrument and example of recorded data.

Due to the ability of detecting shallow object with high resolution, georadar has been applied in many fields with various objectives. Georadar is able to distinguish two different objects based on different electrical properties; hence, georadar are commonly used in various field such as environment study, mining, ground water, ancient artifact, and others. Not only able to detect the electrical properties contrast of material, but also georadar is able to detect the subsurface structure like faults and folds. Hence, the application of georadar for detecting the subsurface structures and monitoring of active faults for mitigation purposes are promising, especially for unstable area in the urban/suburban area with high population where other active source is prohibited.

The exact location of an active fault in urban area is very important to be known for the mitigation purposes. Hence, the potential landslide due to the unstable structure of this area can be warned early to avoid serious hazard or disaster. Many techniques have been used to monitor the stability and mitigate the potential landslide in the area around the active fault which is across the urban area. Nondestructive geophysical methods such as electrical method and electromagnetic method are commonly selected for the investigation and evaluation of the subsurface structure this area. Geo-penetrating radar (GPR) or georadar method is also a common geophysical method that is applied to understand the bedding subsurface and structure in the high-risk area in such condition.

Lembang fault is an example for active fault across urban area with high population density in Bandung, Indonesia. In this area, there are not less than 8 million people leaving around the Lembang fault. The length of this fault itself is about 29 km from east to west part of Bandung [3] as illustrated in **Figure 2**. Because of the compression system in this area, it is predicted that a huge accumulated energy is concentrated in this fault and potentially can be released any time as an earthquake. The earthquake then is predicted also which leads to trigger the local landslide in this area.

Due to some reasons such as soil stability and environmental concern, techniques such as seismic refraction and seismic reflection that use dynamite explosion as a source, are not allowed. Hence the use of georadar technique for identifying

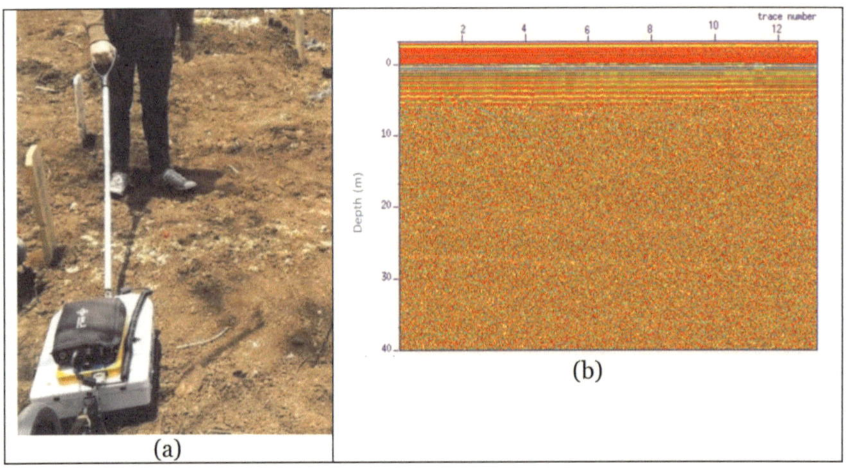

Figure 1.
(a) Georadar instrument [2], (b) example of recorded subsurface data.

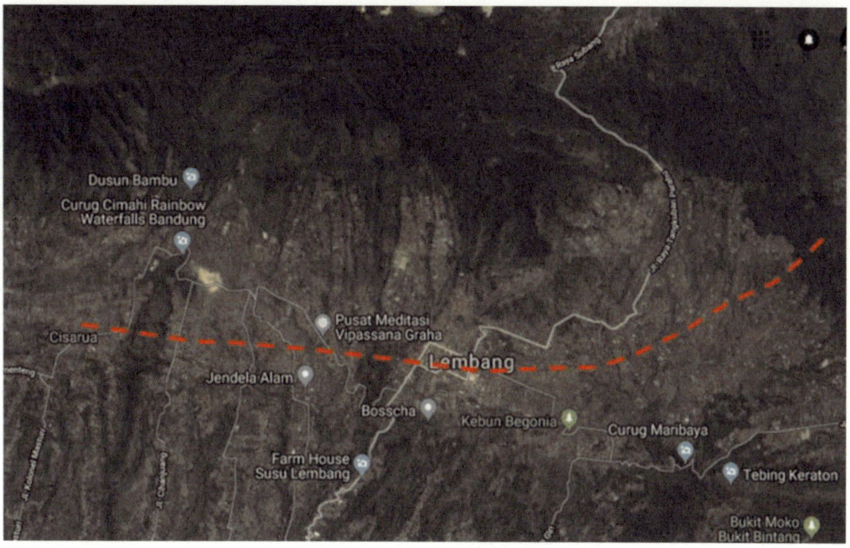

Figure 2.
Active fault (Lembang fault) located from west to east part of Bandung.

fault system in this area becomes more significant. This chapter discusses the example of GPR technique for detecting fault in urban area. The background theory of GPR, design survey, and data gathering, processing, and interpretation of GPR data are discussed and applied for detecting an active fault of Lembang fault in Bandung, Indonesia.

2. Background theory

2.1 Georadar wave propagation

Georadar technique is developed based on electromagnetic wave propagation theory. In one dimension (1D), the propagation of electromagnetic wave in z-direction is explained by Maxwell equation:

$$\frac{\partial^2 E}{\partial z^2} = \mu\varepsilon \frac{\partial^2 E}{\partial t^2} \tag{1}$$

The propagation of the electromagnetic wave is perpendicular to the electrical field (E) and magnetic field (H) and controlled by the velocity and attenuation of medium. The properties of medium are also related to the mineral composition and also water saturation of medium. The velocity of wave propagation in the medium depends on the velocity of electromagnetic wave in the vacuum (c = 0.3 m/ns), relative dielectric constant (ε_r), and relative magnetic permeability ($\mu_r = 1$ for nonmagnetic material):

$$V_m = c/\left\{(\varepsilon_r\mu_r/2)\left[\left(1+P^2\right)+1\right]\right\}^{1/2} \tag{2}$$

where $P = \sigma/\omega\varepsilon$ is an absorption factor and σ is conductivity of medium, $\varpi = 2\pi f$, where f is frequency and $\varepsilon = \varepsilon_r\varepsilon_o$ is vacuum permittivity (8.854×10^{-12} F/m). For the material with low loss, P~ 0, the velocity of georadar is

$$V_m = c/\sqrt{\varepsilon_r} = 0.3/\sqrt{\varepsilon_r}$$

Recording of georadar data is based on the reflection responses of dielectric contrast of medium. If dielectric contrast in the interface between different layers is strong, the reflectors have a strong amplitude in the georadargram. The strength of reflection (reflection coefficient R) is determined by the contrast of velocity and relative dielectric of medium at the boundary. The number of reflected energy is proportional to R:

$$R = \frac{(V_1 - V_2)}{(V_1 + V_2)} \text{ or } R = \frac{\sqrt{\varepsilon_2} - \sqrt{\varepsilon_1}}{\sqrt{\varepsilon_2} - \sqrt{\varepsilon_1}} \tag{3}$$

where v_1 and v_2 are wave velocity of georadar at first and second layers and ε_1 and ε_2 are relative dielectric constant at first and second layers, respectively.

The energy loss during wave propagation is determined by some factors: antenna, transmission between air and soil, reduction due to the configuration or distance between transmitter and receiver, attenuation, and diffraction due to the sharp object. The energy reduction due to the wave propagation between transmitter and receiver is proportional to $1/r^2$, where the r is distance measured between source/antenna and receiver (another antenna), attenuation factor which defend on the dielectric properties of medium, and its magnetic and electrical field itself in the medium. Amplitude will be reduced in the depth of penetration due to the attenuation which is proportional to 1/e (about 37%) of initial energy which also called as skin depth. The skin depth depends on soil resistivity. The ratio of two different amplitudes is formulated as

$$E_o/E_x = \exp(-\alpha x) \tag{4}$$

where α is attenuation coefficient

$$\alpha = \omega\left\{\left(\frac{\mu\varepsilon}{2}\right)\left[\left(1 + \frac{\sigma^2}{\omega^2\varepsilon^2}\right)^{1/2} - 1\right]\right\}^{1/2} \tag{5}$$

Loss factor $(P) = \sigma/\omega\epsilon = \tan D$ and skin depth is defined as

$$\delta = 1/\alpha$$

If D < <1, $\delta = (2/\sigma)(\varepsilon/\mu)^{1/2}$, the skin depth is formulated:

$$\delta = (5.31\sqrt{\varepsilon_r})/\sigma$$

where σ is an electrical conductivity (mS/m).

In the saturated porous medium, the loss energy is proportional to the conductivity and invers proportionally to the relative dielectric constant and frequency. The conductivity and relative dielectric constant is dominated by fluid saturant compared to the matrix itself. The bulk of relative dielectric constant (ε_r) is roughly equal to porosity and relative dielectric constant of fluid (ε_r). Due to the fluid/saturant is more conductive, the attenuation becomes higher. The geological material mostly have a relative dielectrical constant is about 3–30, and hence the georadar wave velocity is about 0.06–0.175 m/ns, while the air velocity is 299.8 mm/ns.

The relationship between dielectric, permittivity, and conductivity of medium is governed by

$$\varepsilon^* = \varepsilon' + i\left(\varepsilon'' + \sigma_s/\omega\varepsilon_o\right) \tag{6}$$

where ε^* is a complex permittivity, ε'' is an imaginary part of permittivity, and σ_s is static or DC conductivity, while the complex conductivity is

$$\sigma^* = \sigma' + i\sigma'' = j\omega\varepsilon_o\varepsilon^* \tag{7}$$

where ω is angular frequency.

In the porous medium material where water is a saturant, the bulk dielectric constant and porosity (ϕ) is defined as

$$\varepsilon_r = (1 - \phi)\varepsilon_m + \phi\varepsilon_w \tag{8}$$

where ε_m and ε_w are dielectric constant of matrix and water, respectively. By using a simple relation, $V = c/\sqrt{\varepsilon_r}$, for low loss material where c is georadar wave velocity in the air, the velocity of georadar in the medium is defined as

$$V = c/[(1 - \phi)\varepsilon_m + \phi\varepsilon_w] \tag{9}$$

Eq. (9) shows that if velocity can be extracted, then porosity of medium can be predicted or vice versa.

2.2 Configuration and resolution of georadar survey

Effectiveness and recoverable of georadar survey is determined by the configuration of the survey. At least there are two types of acquisition: monostatic mode and bistatic mode. In the monostatic mode, one antenna is used as transmitter and receiver simultaneously. While in the bistatic mode, the receiver and transmitter are separated using different antenna. Based on the target itself, the configuration of data acquisition can be performed using different ways: radar reflection profiling, wide-angle reflection and reflection (WARR) and common midpoint (CMP) sounding, and transillumination or tomography (**Figure 3**).

The vertical resolution of georadar is defined by its frequency or wavelength. Each antenna of georadar is designed for certain frequency range, where the peak energy will be associated in the peak frequency of the signal. Hence the vertical resolution georadar signal is determined by wavelength divide by four ($\lambda/4$). Meanwhile, the horizontal resolution of georadar is controlled mainly by the number of traces/s (or traces/m), the beam width, the radar cross section of the reflector, and the depth where target is located [4]. The conical beam of georadar signal itself is inversely proportional to the square root of attenuation coefficient ($\sqrt{\alpha}$). It means that the horizontal resolution is better in the medium with high attenuation coefficient [5].

2.3 Data processing

Data processing in radargram depends on the objectives; there is no standard processing workflow. However, usually the processing data is done to gain the signal which is attenuated during propagation, removing some noise by filtering, deconvolution, and diffraction reduction through migration process. For certain purposes, sometimes the conversion from time domain needs to be done to get the depth domain; in this case the velocity model is needed.

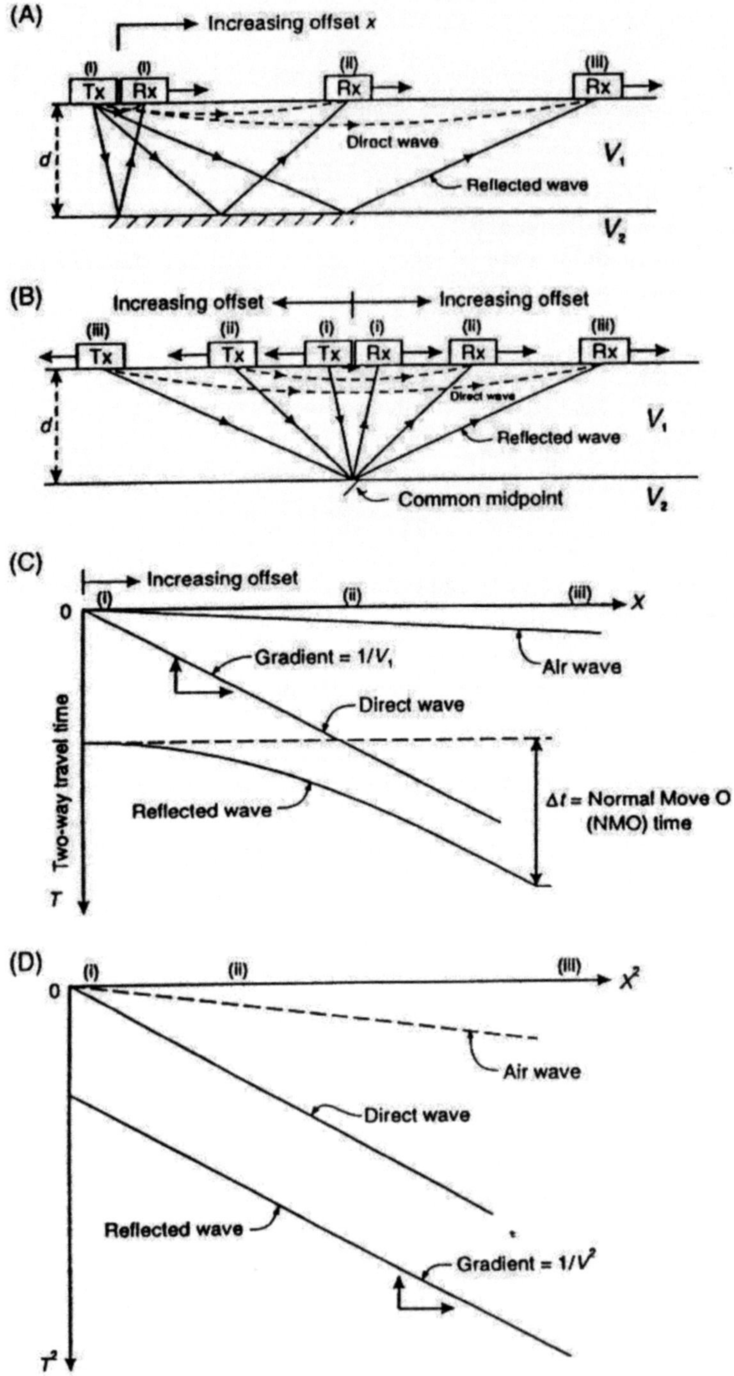

Figure 3.
(A) Sounding WARR dan, (B) sounding CMP, (C) graph of time-offset (T-X) with NMO, (D) graph of relationship T2–X2 [3].

Gain is performed to amplify the amplitude decay due to the distance of propagation. Factors affecting the amplitude decay are attenuation and spherical divergence propagation. Gain is performed by applying a gain function g(t):

$$\text{Gain (dB)} = A.t + B.20\ \log(t) + C \tag{10}$$

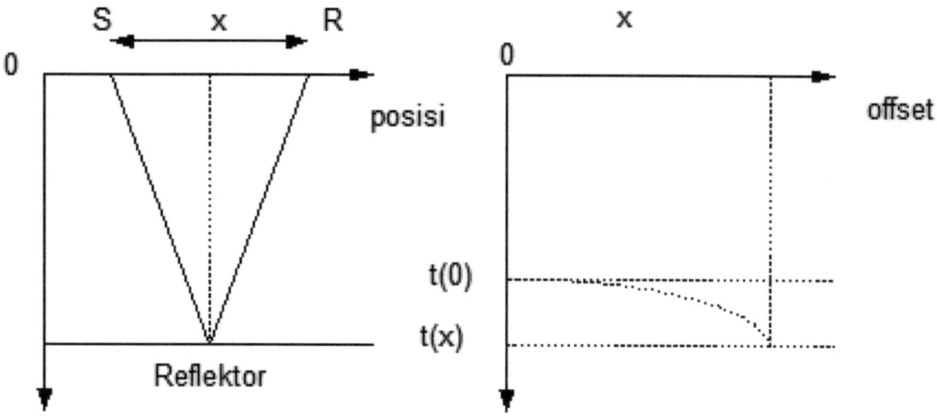

Figure 4.
Schematic of transmitter and receiver related to the NMO process.

where t is travel time, A is attenuation factor, B is spherical divergence factor, and C is gain constant.

In the implementation on processing, programmed gain control (PGC) and automated gain control (AGC) are commonly used. In the PGC, the gain function is estimated by interpolating the amplitude at certain window sample. While in the AGC, the gain function is generated by taking root mean square (RMS) on each amplitude at certain window. The gain function g(t) is interpolated in the central of the selected window. The window length selection affects the reflector strength in the result. If the window length is too wide, the signal from deeper part will gain less, and if selected window is too small, all the reflection will be gained strongly; in this case it is difficult to distinguish strong reflector from others because all reflector will be gained strongly.

The velocity analysis needs to be conducted to know the value of velocity (velocity model); hence, the true depth and slope can be estimated. Direct measurement of medium velocity can be done from wellbore or indirectly through velocity analysis during normal moveout (NMO) process. The NMO process is illustrated in **Figure 4**:

The function of NMO is describes as follows:

$$T = \frac{2S}{V} = \frac{2\sqrt{\left(\frac{x}{2}\right)^2 + h^2}}{V} \tag{11}$$

$$T^2 = \frac{4\left(\frac{x^2}{4} + h^2\right)}{V^2} = \frac{x^2}{V^2} + \frac{4h^2}{V^2}, \text{ if } \mathrm{T}^2(0) = 4\mathrm{h}^2/\mathrm{V}^2$$

$$\text{Then } T^2(x) = T^2(0) + \frac{x^2}{V^2}$$

where V is velocity obtained from relation of reflection time at zero offset and offset and distance/offset.

3. Acquisition and processing data

Geological of Bandung area is still young and renewed because of volcanic activities around Bandung area. Based on previous study, there is a Lembang fault in Bandung area which occurred from tectonic process. This fault located in the southern part of Lembang and crossing Cisarua from east to west of Manglayang mountain. Throw of this fault is varying up to 450 m in Pulasari near to the target area. The Lembang fault was created during the Pleistosen era (about 500,000 years ago) [6].

In the end of Miocene, series of mountains and folds are created in the northern part and in the southern part which become series of volcanos. In the breaking time of Pliocene era, there is no activity of volcanoes and sedimentation, and in the end of Pliocene era, series of mountains were created, and sediments in the northern part were folded and shifted into northern part of Bandung. Materials as a result of volcano eruption activities are distributed into southern part of Bandung.

To detect the existence of Lembang fault, a georadar survey was conducted in this area. Sketch of data gathering in this area is shown in **Figure 5**. The data collection was conducted using common-offset method with 25 and 50 MHz antenna. Most of the profile was selected perpendicular to the fault.

All the data are processed using Seismic Unix (SU) software by performing the scaling on the time sampling rate from nanosecond to millisecond (ms), frequency from megahertz to hertz and velocity from m/ns to m/μs. The details of conversion factor are shown in **Table 1**.

The processing data includes filtering using band-pass filter (10, 20, 30, 50) MHz for antenna 25 MHz and (10, 30, 70, 100) MHz for 50 Hz, AGC, and velocity analysis based on hyperbole fitting curve. The velocity obtained from hyperbole fitting curve is shown in **Table 2**. This velocity model was used to convert time

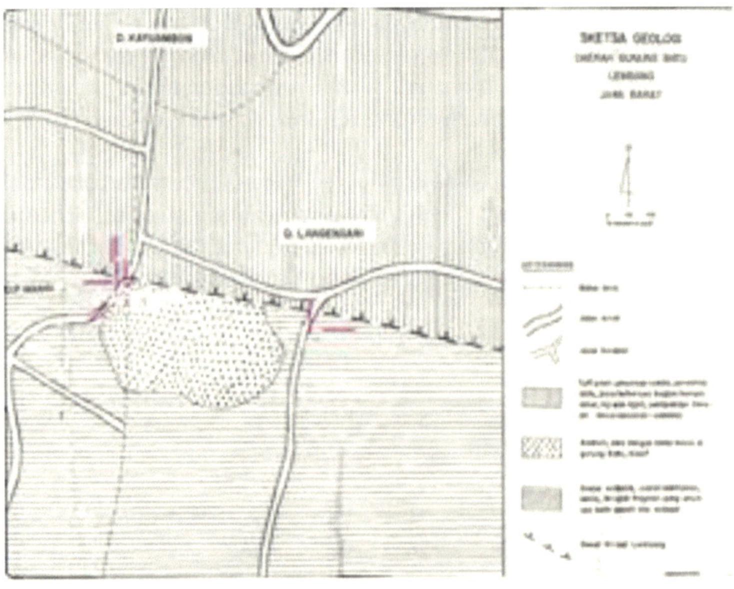

Figure 5.
Georadar data acquisition map.

Parameters	True value	Scaled value
Time sampling rate	1.4 ns	1.4 ms
Nyquist frequency	357 MHz	357 Hz
Offset	1 m	1 m
Radar speed in air	0.3 m/ns	300 m/μ

Table 1.
Conversion from radar into seismic scale.

Depth (m)	Velocity (m/μs)
15.63	196.17
17.16	186.37
22.25	180.69

Table 2.
Velocity model extracted from hyperbole fitting curve during velocity analysis.

domain into depth domain in the profile. In general, the velocity is quite high because the lithology is dominated by tuff, andesite, and breccia volcanic.

4. Result and discussion

Out of several profiles which were studied, profile 03 displayed the large fault in the radargram which is associated with a main Lembang fault. In the other profiles, there are some small fault systems. Based on the velocity analysis, the structure of Lembang fault is a conductive area where the velocity decreases with depth. Profile in **Figure 6** which is taken perpendicular to the fault shows a normal fault system. The foot wall is located in the northern part and hanging wall in the southern part. The position of foot wall part is lower about 7–8 m compared to the hanging wall part. The structure of this area consists of basement which is indicated by a free reflection area and sediment bedding in the horizontal layer. Above the foot wall part, there is a pattern of unconformity.

In **Figure 6**, the top of basement formation is interpreted as the blue line indicates a normal fault. The folded reflection can be resulted as post fault due to the compression from the northern part; the horizontal bedding is folded as small anticline. Cracking in the shoulder of road around this profile indicates that the fault reaches the surface. Those crack lies in the west–east direction where the northern part is lower than southern part. Previous study mentioned that this active

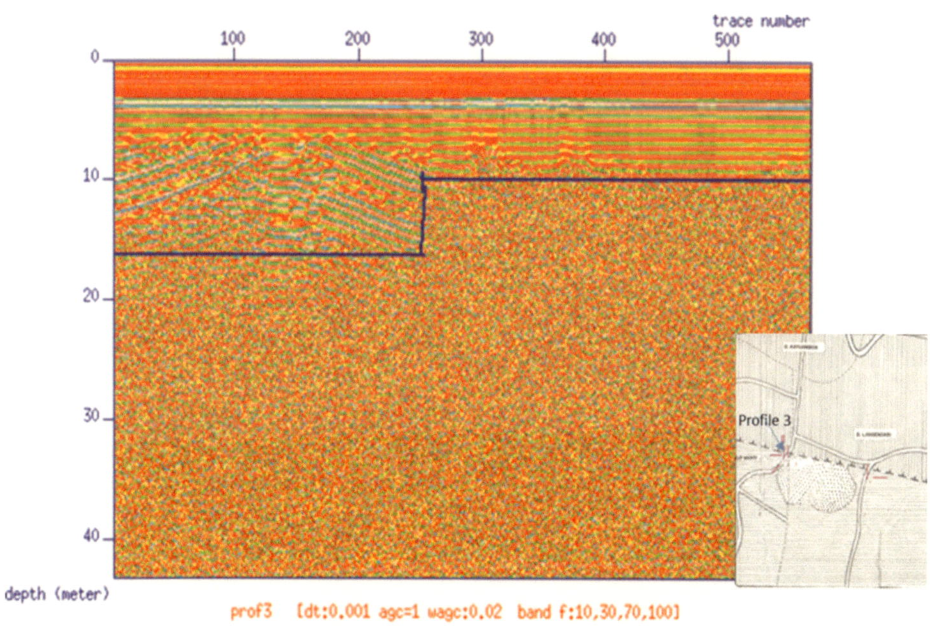

Figure 6.
Cross section taken perpendicular to the fault.

fault has a movement rate per year about 0.3–1.4 cm/year [6]. Instability of this area due to the activities of this fault especially the possible earthquake needs to be monitored further to avoid the further effect like landslide which can damage the urban and suburban area around this fault.

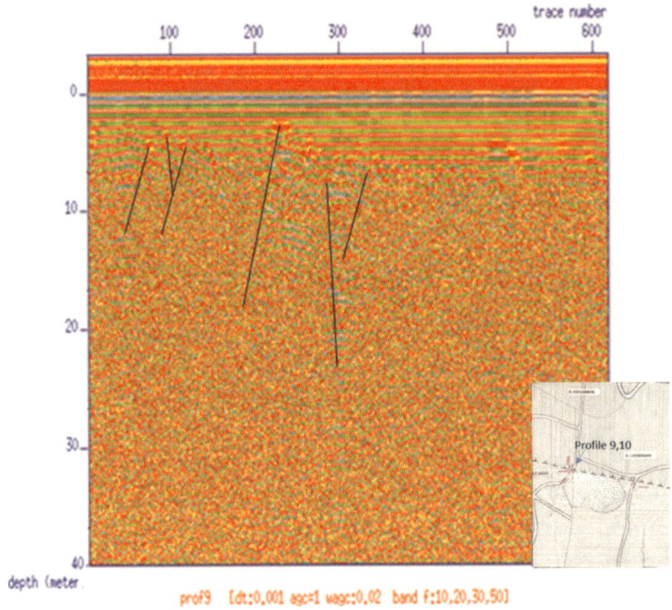

Figure 7.
Profile 9 taken perpendicular to the fault line.

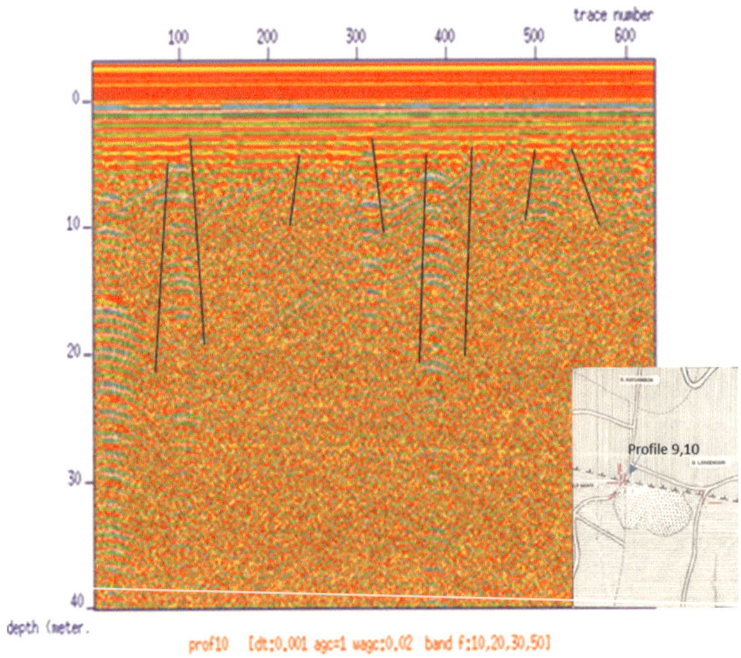

Figure 8.
Profile 10 taken perpendicular to the fault direction.

Small faults around the main fault also recorded in other profiles. **Figures** 7 and 8 show the pattern of small fault systems which are still related to the activity of the main fault. Profiles 9 and 10 as shown in **Figures 4** and **5** are taken perpendicular also to the main fault direction. The main fault is not recorded in these profiles. However, the diffraction pattern which indicates small fault system is appearing in this area.

The subduction and compression process in the north–south direction also produces other local fault system. Because the length of this fault is only 29 km, the maximum earthquake due to the energy release in this area is predicted that the earthquake magnitude will not be more than 6 in Richter scale. Tectonic activity record in this area showed that the focus of earthquake is located in the depth of 3–7 km. The earthquake activities are also related to the continuity of three main fault in Bandung area which are Cimandiri fault, Lembang fault, and Baribis fault [6]. Even though the earthquake recorded in this area are not strong earthquakes, because this area is one of the tourism object locations in Bandung and high population around this area, the monitoring of possible hazard needs to be continued. A small earthquake is possible to activate that fault which can trigger instability of the soil mechanism in this area.

5. Conclusion

Lembang fault which is located near the urban area with high population density is successfully imaged using georadar method. Based on radargram result, the foot wall part of Lembang fault is located in the northern part, and hanging wall is located in the southern part. The subsurface structure in this area is dominated by basement and sediment layers. Due to the location of this fault which is near the urban area with high population, further investigation to mitigate the potential of landslide, instability, and activity of this fault needs to be monitored. A CMP survey type can be proposed to be used to improve velocity information; hence, the depth of structure in the subsurface can be improved. A monitoring on the movement of this fault activity using GPS needs to be performed to monitor these activities consciously.

Author details

Maman Hermana[1*], Maya Genisa[2], Luluan A. Lubis[1] and Chow Weng Sum[1]

1 University Technology of PETRONAS, Perak, Malaysia

2 YARSI University, Jakarta, Indonesia

*Address all correspondence to: maman.hermana@utp.edu.my

References

[1] Reynolds JM. An Introduction to Applied and Environmental Geophysics. 2nd ed. Wiley-Blackwell; 2011. Retrieved from http://eu.wiley.com/WileyCDA/WileyTitle/productCd-0471485357.html

[2] https://research.lppm.itb.ac.id/2017/07/04/detection-of-burried-bodies-on-landslide-area-using-ground-penetra ting-radar-method/

[3] Ramadhan ML, Prawita SM, Fatmasari NW. Identifikasi Bidang Patahan Sesar Lembang dengan Metode Electrical ResistivityTomography untuk Mitigasi Bencana Gempa Bumi dan Longsor. In: Geostone UPN Conference. 2016

[4] Rial FI, Pereira M, Lorenzo H, Arias P, Novo A. Vertical and horizontal resolution of GPR bow-tie antennas. In: Proceedings of the 2007 4th International Workshop on Advanced Ground Penetrating Radar, IWAGPR; July 2007. pp. 187-191. DOI: 10.1109/AGPR.2007.386549

[5] Daniels DJ, Gunton DJ, Scott HF. Introduction to subsurface radar. IEE Proceedings F Communications, Radar and Signal Processing. 1988;**135**(4):278. DOI: 10.1049/ip-f-1.1988.0038

[6] Firdaus MW, Setyawan A, Yusuf M. Gayaberat second vertical gradient studi kasus sesar. Younster Physics Journal. 2016;**5**(1):21-26

Chapter 5

Dramatic Weakening and Embrittlement of Intact Hard Rocks in the Earth's Crust at Seismic Depths as a Cause of Shallow Earthquakes

Boris Tarasov

Abstract

Frictional stick-slip instability on pre-existing faults is well studied experimentally and considered as the general mechanism for shallow earthquakes. At the same time, post-peak properties of intact hard rocks under high confining stresses σ_3 corresponding to seismic depths of shallow earthquakes are still unexplored experimentally due to uncontrollable and violent failure of rock specimens even on modern stiff and servocontrolled testing machines. The lack of knowledge about post-peak properties of the majority of the earthquake host rocks prevents us from understanding and quantifying the contribution of these rocks to shallow earthquakes. This paper discusses a recently identified shear rupture mechanism operating in hard rocks under high σ_3 which causes dramatic rock weakening and embrittlement (by tenths of times) during the post-peak failure. The unknown before 'abnormal' properties of hard rocks imply the fundamentally different general mechanisms for shallow earthquakes. It is shown that in the earth's crust, the new mechanism acts in the vicinity of pre-existing faults only and provides the formation of new dynamic faults in intact rocks at very low shear stresses (significantly less than the frictional strength). The fault propagation is characterised by extremely low rupture energy and small stress drop. These 'abnormal' properties make hard rocks the main and more dangerous source of shallow earthquakes in comparison with pre-existing faults.

Keywords: extreme shear ruptures, shear rupture mechanism, hard rocks, shallow earthquakes, rock weakening and embrittlement

1. Introduction

Today, frictional stick-slip instability on pre-existing faults is considered as the general mechanism of shallow earthquakes. This concept is based on a number of features observed for natural and laboratory earthquakes which support the stick-slip nature of earthquakes as opposed to failure of intact rocks. Some of these features are: nucleation of earthquakes at tectonic plate boundaries and other pre-existing faults, recurring stick-slip instability along a pre-existing fault, low shear stresses activating earthquakes, small stress drop and specific depth-frequency distribution of earthquake hypocentres with a maximum at a special depth [1–9].

The nature of friction and stick-slip instability for rocks in association with earthquakes has been comprehensively studied during the last half of a century [1, 5, 7, 10–15]. At the same time, post-peak properties of intact hard rocks under high confining stresses σ_3 corresponding to seismic depths of shallow earthquakes are still unexplored experimentally. The reason for that is uncontrollable and violent failure of rock specimens even on modern stiff and servocontrolled testing machines. Today, post-peak properties of hard rocks at high σ_3 and the failure mechanism operating at these conditions are treated by analogy with conventional understanding based on experimental results obtained for softer rocks. The paper shows that this analogy is unacceptable because the real properties of hard rocks at high σ_3 differ fundamentally from the conventional understanding. The lack of knowledge about post-peak properties of the majority of the earthquake host rocks prevents us from understanding and quantifying the contribution of these rocks to shallow earthquakes.

The paper discusses a recently identified shear rupture mechanism operating in hard rocks under high σ_3 that is responsible for extreme rupture dynamics [16–23]. The mechanism was identified on the basis of comprehensive analysis of side effects accompanying extreme ruptures. The new mechanism provides two remarkable features in the rupture head: (1) low shear resistance approaching zero and (2) highly amplified shear stresses. The combination of these features allows for a shear rupture to propagate through intact rock spontaneously at very low shear stresses applied with the absorption of a small amount of energy which indicates dramatic rock weakening and embrittlement at high σ_3. Due to the weakening and embrittlement during the failure process, the stress-strain curves for laboratory specimens look very specific in the post-peak region indicating 'abnormal' properties.

The paper demonstrates that in the earth's crust, the new mechanism is active in the vicinity of pre-existing faults only and can provide the formation of new dynamic faults in intact rocks at very low shear stresses (up to an order of magnitude less than the frictional strength). The fault propagation is accompanied by extremely low rupture energy and small stress drop which can be smaller than for stick-slip instability on pre-existing faults. It is shown that some earthquake features currently attributed to the stick-slip instability on pre-existing faults can be provided by the new mechanism for ruptures propagating in intact rocks. Some of these features are nucleation on the basis of pre-existing faults, recurring instability on a pre-existing fault, activation of earthquakes at low shear stresses, small stress drop and specific depth-frequency distribution of hypocentres.

The unknown before 'abnormal' properties of hard rocks at high σ_3 make intact rocks more dangerous in respect of shallow earthquakes than pre-existing faults because the new mechanism can generate dynamic faults at shear stresses significantly below the frictional strength. The proximity of the pre-existing fault to the zone of dynamic new fracture development in intact rock creates the illusion of frictional stick-slip instability on the pre-existing fault, thus concealing the real situation. According to the new knowledge, intact hard rocks adjoining pre-existing faults represent the general source of shallow earthquakes.

2. Post-peak properties of hard rocks at high σ_3: conventional and new understanding

Figure 1 illustrates the situation with our current knowledge about rock properties beyond the peak stress at triaxial compression of type $\sigma_1 > \sigma_2 = \sigma_3$. Such

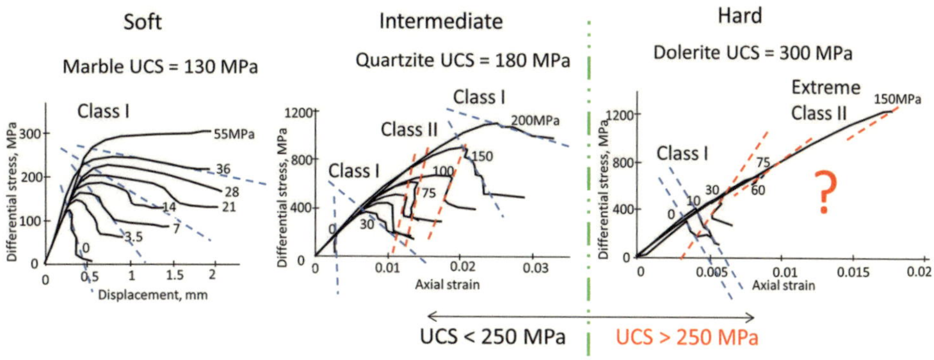

Figure 1.
Three sets of stress-strain (displacement) curves illustrating features of post-peak behaviour for soft, intermediate and hard rocks.

stress conditions are normally used in laboratory experiments to simulate on rock specimens stress conditions typical for the earth's crust at different depths. **Figure 1** shows three sets of stress-displacement curves obtained at different levels of confining stress σ_3 for rocks of different hardness. Rock hardness here is characterised by uniaxial compressive strength (UCS) and increases from left to right. The curves demonstrate how rock hardness affects post-peak behaviour at rising σ_3. Relatively soft rocks on the left (represented by marble with UCS = 130 MPa [24]) exhibit class I behaviour at all levels of σ_3 (indicated by blue dotted lines) and show an increase in post-peak ductility with rising σ_3. Rocks of intermediate hardness (represented by quartzite with UCS = 180 MPa) exhibit post-peak embrittlement within a certain range of σ_3 which is expressed by transition from class I to class II behaviour (indicated by red dotted lines). At lower and higher σ_3, the post-peak ductility increases with rising σ_3.

The typical behaviour of hard rocks (represented by dolerite with UCS = 300 MPa) is characterised by dramatic embrittlement at high σ_3 leading to extreme class II behaviour. It is important to note that the transition from class I to class II for this rock occurs at σ_3 = 30 MPa. At relatively low σ_3, the post-peak failure can be controlled both for class I and class II on stiff and servocontrolled testing machines. However, at σ_3 > 50 MPa, the failure process associated with propagation of a shear rupture becomes uncontrollable and abnormally violent. With rising σ_3 rock brittleness and the violence increase. However, we can suppose that at very high σ_3, by analogy with intermediate rocks, hard rocks should also return to more ductile behaviour. Hard rocks with similar post-peak behaviour are represented mainly by volcanic and highly metamorphic rocks with UCS > 250 MPa.

Figure 1 demonstrates that the effect of embrittlement within a certain range of high σ_3 (different for different rocks) increases with increasing rock hardness. The problem is that all existing ultra-stiff servocontrolled testing machines cannot provide controllable failure for hard rocks at high σ_3 corresponding to stress conditions of seismic depth for shallow earthquakes. Due to this limitation in testing capability, post-peak properties of hard rocks at high σ_3 are experimentally unexplored. Today, post-peak properties of hard rocks at high σ_3 and the failure mechanism operating at these conditions are treated by analogy with conventional understanding based on experimental results obtained for softer rocks.

On the basis of comprehensive analysis of side effects accompanying extreme ruptures, an unknown before shear rupture mechanism and inaccessible post-peak properties of hard rocks generated by this mechanism were recently identified [16–23]. In this section we will demonstrate only the fundamental difference

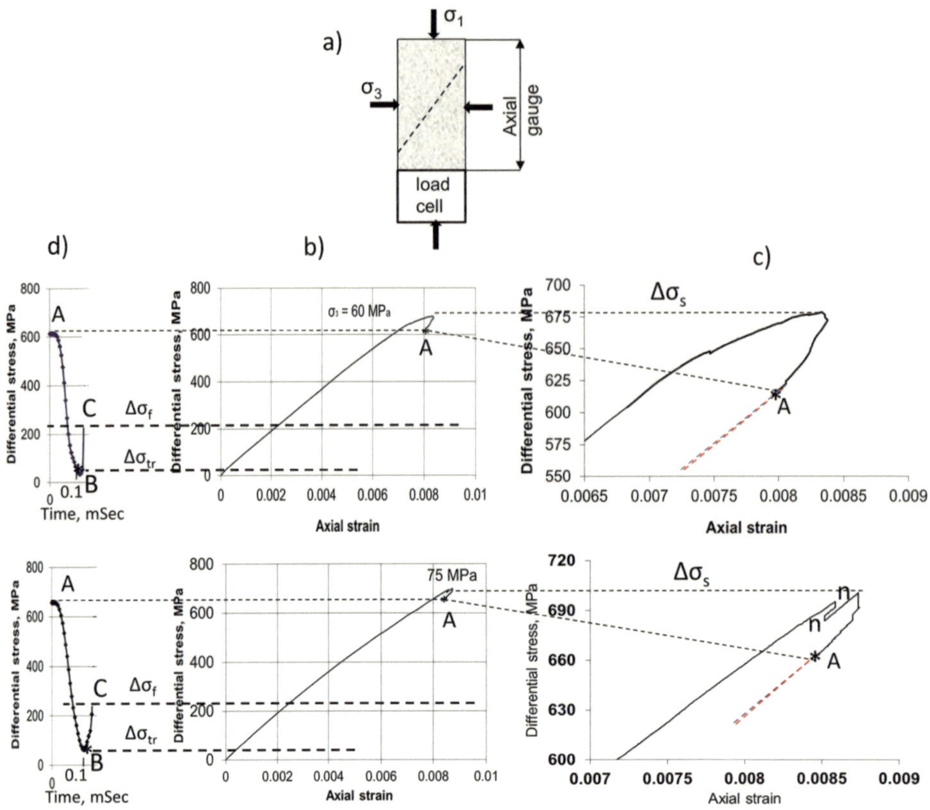

Figure 2.
Some typical features of post-peak behaviour of hard rocks at highly confined compression illustrated by experimental results obtained for dolerite specimens at σ_3 = 60 MPa and σ_3 = 75 MPa.

between the conventional and new understanding of post-peak properties of hard rocks at high σ_3. The new failure mechanism and more detailed information about hard rock behaviour in laboratory and natural conditions will be discussed in the following sections of the paper.

Figure 2 demonstrates some features typical for post-peak behaviour of hard rocks at high σ_3. It shows experimental results obtained for dolerite specimens with UCS = 300 MPa tested under σ_3 = 60 MPa and σ_3 = 75 MPa. **Figure 2a** illustrates schematically a cylindrical specimen equipped with an axial gauge and a load cell as used in experiments. The failure process at high σ_3 is always associated with a shear rupture propagation shown by a dotted line on the specimen body. **Figure 2b** shows two stress-strain curves where points A indicate the final stage of controllable post-peak failure after which the spontaneous and violent shear rupture propagation took place. **Figure 2c** shows enlarged fragments of the stress-strain curves involving the post-peak stage. The post-peak curves here reflecting real rock properties up to point A were easily obtained in static regime due to controllable reverse deformation of the specimen provided by the servo-controlled system. However, beyond point A, the failure control became impossible. The reason for that is as follows. At point A, the post-peak modulus (represented by a red line on the graphs) became practically equal to the unloading elastic modulus (represented by a blue line on the graphs) which corresponds to the extreme Class II behaviour. The unloading modulus was determined by unloading the specimen tested under σ_3 = 75 MPa at the peak stress (marked as n-n on the graph). The fact that the post-peak modulus and unloading modulus are practically coincide indicates that the post-peak rupture energy beyond points

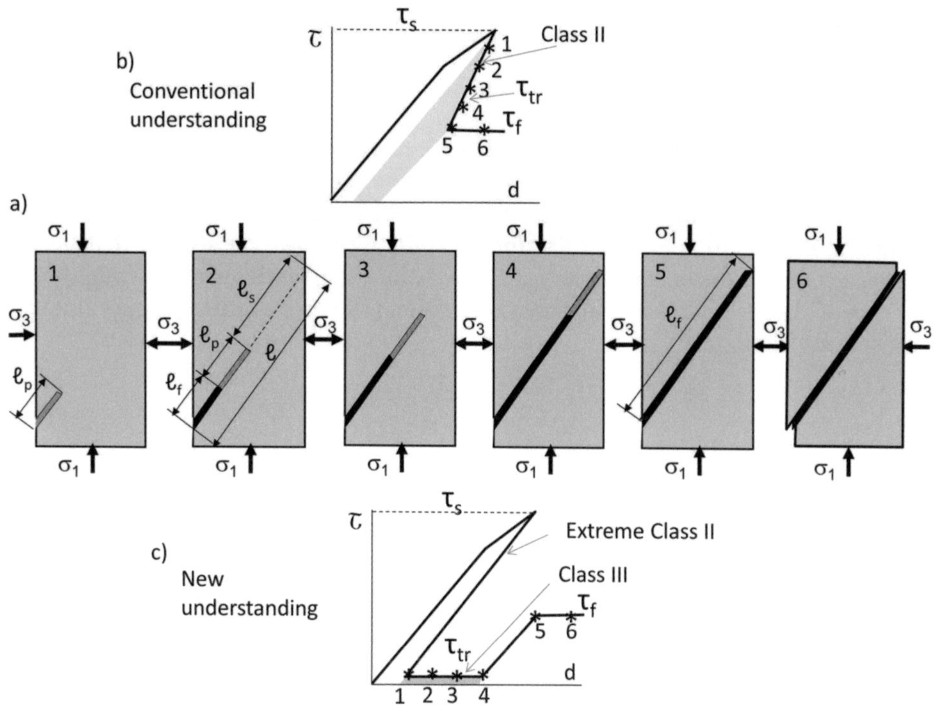

Figure 3.
Schematic illustration of the fundamental difference between the conventional and the new understanding of post-peak properties of hard rocks tested at high σ_3.

A is vanishingly small. The controllable failure at this situation becomes impossible because the existing testing machines unable to provide sufficiently fast the specimen unloading to stop the rupture propagation. The spontaneous failure at high σ_3 is usually abnormally violent and accompanied by strong sound and shudder of the testing machine.

Because during the spontaneous failure the testing machine cannot provide sufficiently fast unloading of the specimen to follow the actual post-peak modulus, the readings of gauges obtained at this stage of failure do not reflect the actual post-peak properties of the failing specimen. At the same time, the conducted above analysis of the post-peak modulus at point A allows supposing that the stress-strain curve beyond point A should be very close (practically coincide) to the unloading elastic curve. Another very important post-peak feature typical for hard rocks at high σ_3 can be observed on the stress-time curves recorded by the load cell in **Figure 2d**. These curves demonstrate that during the spontaneous failure the specimen strength at a certain stage becomes significantly below the static frictional strength, the level of which is represented by a horizontal dotted line $\Delta\sigma_f$. The static frictional strength was determined experimentally by deforming the failed specimen at low strain rates. The least level of the specimen transient strength during the spontaneous failure corresponding to point B is indicated by a horizontal dotted line $\Delta\sigma_{tr}$ that is about 10 times less than the static frictional strength $\Delta\sigma_f$. It will be shown later that the observed in these experiments the extreme Class II behaviour and the abnormally low specimen transient strength are provided by the new shear rupture mechanism.

Using the obtained results, we can formulate a hypothesis about the fundamental difference in post-peak properties of hard rocks at high σ_3 in terms of the conventional and the new understanding illustrated in **Figure 3**. **Figure 3a** shows six stages of shear rupture propagation through a specimen. The rupture incorporates a process zone ℓ_p representing the rupture head and a frictional zone ℓ_f located behind the head. In intact zone ℓ_s, located in front of the process zone, shear resistance corresponds to the intact material strength τ_s, while behind the process zone (in zone ℓ_f), shear resistance is equal to the frictional strength τ_f. After completion of the process zone at stage 1, the length of ℓ_p stays constant, while the length of the frictional zone ℓ_f increases during the rupture propagation.

In accordance with conventional understanding, the specimen strength beyond the peak stress (transient strength τ_{tr}) at any failure stage is determined roughly by the sum of shear resistance of all three zones along the propagating rupture (intact, process and frictional):

$$\tau_{tr} = \tau_s \frac{l_s}{l} + \tau_p \frac{l_p}{l} + \tau_f \frac{l_f}{l} \quad (1)$$

Shear resistance of the process zone here can be determined roughly as $\tau_p = (\tau_s + \tau_f)/2$.

Figure 3b illustrates the conventional understanding of post-peak properties of hard rocks at high σ_3. Points on the graph indicate six identical failure stages shown for the specimen in **Figure 3a**. The specimen strength at each stage is described by Eq. (1). Here, during the failure process, the transient strength decreases gradually by substituting the material strength with frictional strength. At stages 5 and 6, the specimen strength is determined by friction in the fault which is considered today to be the lower limit on rock shear strength. The post-peak rupture energy between stages 1 and 5 corresponds to the shaded area under the curve.

Figure 3c illustrates the new understanding of post-peak properties of hard rocks at high σ_3. It will be shown later that the new mechanism provides very low shear resistance of the completed process zone which can be $\tau_p \approx 0.1\tau_f$. Furthermore, the new mechanism represents a very powerful stress amplifier (based on an unknown before principle) providing high shear stresses in the process zone at low shear stresses applied. These two unique features after completion of the process zone at stage 1 make it possible for the shear rupture to propagate through intact rock even at very low shear stresses applied τ which can be significantly (up to an order of magnitude) less than the frictional strength. In this case the specimen transient strength at controllable failure is determined solely by the process zone strength: $\tau_{tr} = \tau_p$.

Figure 3c shows that controllable failure can be provided if the testing machine is capable to unload the specimen up to the level $\tau_{tr} = \tau_p$ at the moment of completion of the process zone (stage 1). The extreme Class II stage beyond the peak stress is associated with the initial formation of the process zone. After that the rupture can propagate statically through the specimen at applied stresses slightly above $\tau_{tr} = \tau_p$ which represents the specimen strength between failure stages 1 and 4. The rupture propagation through intact rock at a constant shear stress significantly below the frictional strength we will classify as Class III. The post-peak rupture energy between stages 1 and 4 corresponds to the shaded area on the graph. This very low energy absorption implies very high brittleness of the material at the failure process. After stage 5 when the process zone has crossed the specimen, the situation changes. The specimen strength at stages 5 and 6 is determined by friction in the developed fault. To provide displacement along the developed fault, it is necessary to increase the applied stress up to τ_f.

It should be noted that if the testing machine cannot provide sufficiently fast unloading at the moment of completion of the process zone (stage 1), the failure process will be spontaneous and violent because in this case the applied stress

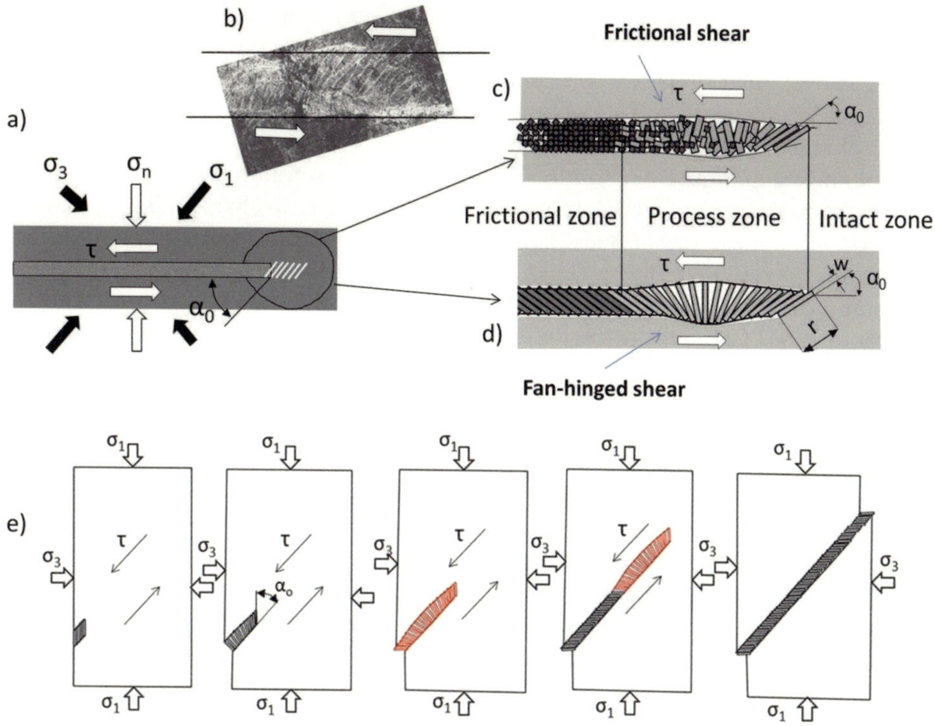

Figure 4.
a) and b) Nature of shear rupture growth in hard rocks at high σ_3. c) and d) The difference between the conventional (frictional) and the new (fan-hinged) shear rupture mechanisms. e) Illustration of the fan structure formation and propagation through a rock specimen.

exceeds the material strength which corresponds to $\tau_{tr} = \tau_p$. It is important to note also that in the case of spontaneous failure, the load cell adjoining the specimen (see **Figure 3a**) will record the variation of stresses applied to the specimen from the testing machine but not the actual material strength determined by the process zone. The actual strength of the process zone can be reflected by the load cell when the applied stress decreases to the level $\tau_{tr} = \tau_p$ which corresponds to point B in **Figure 2d**. The new shear rupture mechanism that is responsible for the discussed 'abnormal' post-peak properties including Class III behaviour of hard rocks under high σ_3 will be introduced in the next sections.

3. General principles of the new mechanism operating in hard rocks at high σ_3

3.1 Structure of shear ruptures

Figure 4 illustrates the nature of shear rupture propagation in brittle intact rocks at high σ_3. In **Figure 4a** a shear rupture propagates from left to right under stresses σ_1, σ_3, σ_n and τ representing the applied major and minor stresses and the induced normal and shear stresses. Shear ruptures are known to propagate through brittle rocks because of the creation of an echelon of tensile cracks at the rupture tip generated along the major stress that is at angle $\alpha_o \approx (30° \div 40°)$ to the shear rupture plane [25–28]. The echelon of inclined tensile cracks and inter-crack slabs forms a typical structure of shear ruptures illustrated by a photograph in **Figure 4b** (modified from [29]). Horizontal lines here indicate the rupture faces. It was observed that at relative displacement of the rupture faces, inter-crack slabs are subject to rotation [25–28]. We will call hereafter the inter-crack slabs as domino blocks.

Figure 4c and **d** show two fundamentally different behaviours of domino blocks at rotation which characterise the conventional and the new understanding of failure mechanisms operating in hard rocks at high σ_3. According to the conventional understanding in **Figure 4c**, domino blocks collapse at rotation creating friction within the process zone in the rupture head [25–28]. This mechanism is associated with frictional shear and provides conventional post-peak properties illustrated in **Figure 3b**. According to the new understanding (**Figure 4d**), domino blocks can withstand the rotation without collapse at a certain combination of such parameters as domino block geometry (ratio between the block length r and width w), rock strength (determining the strength of domino blocks) and applied stresses. Because the relative shear displacement of the rupture faces increases with distance from the rupture tip, the successively generated and rotated domino blocks form a fan-like structure that represents the rupture head [16–23]. This mechanism is associated with fan-hinged shear within the fan zone and with the following two fantastic features of the fan structure: (1) extremely low shear resistance approaching zero and (2) high amplification of shear stresses.

The fan mechanism is responsible for the 'abnormal' post-peak properties of rock specimens illustrated in **Figure 3c. Figure 4e** shows different stages of the fan-structure formation and propagation on the basis of tensile cracking process in a rock specimen. The orientation of tensile cracks and domino blocks in the propagating rupture tip is along σ_1. However, due to relative displacement (shear) of the rupture faces, the domino blocks behind the tip consistently rotate and form the fan structure. Next sections considering unique features of the fan structure will demonstrate that after completion of the fan structure and up to the moment at which the fan has crossed the specimen body, the specimen strength is very low and corresponds to stages 1–4 in **Figure 3c**.

3.2 Physical model of the fan mechanism

The problem is that the direct experimental study of the fan mechanism is impossible today, firstly, because modern stiff and servocontrolled testing machines do not allow stopping the failure process governed by the fan mechanism and, secondly, because the fan structure is a transient phenomenon. The fan structure can be in the stable condition during its initial formation before the instability starts and at the final stage of the rupture termination. However,

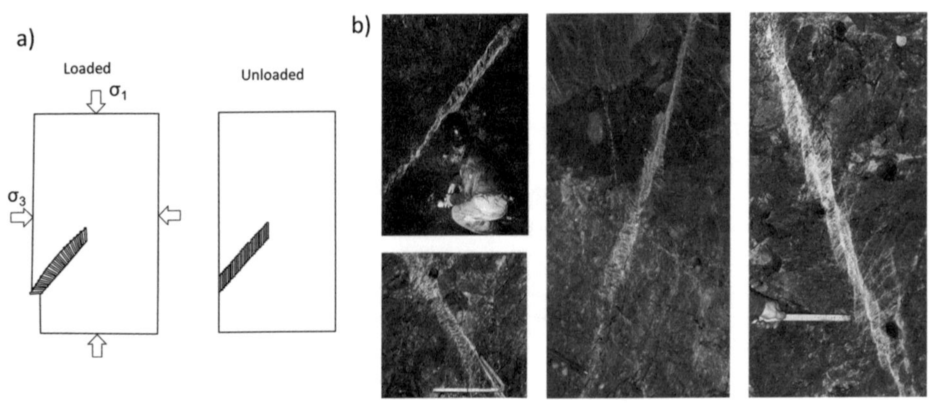

Figure 5.
a) Transient nature of the fan structure and b) the conventional domino-like structure of shear ruptures.

in both cases during total unloading (e.g. associated with the removal of stresses from the specimen or with the tectonic exhumation of the rock mass involving the fan structure), reverse elastic deformations transform the fan into the conventional domino-block structure. **Figure 5a** illustrates this situation schematically on the laboratory specimen. The left specimen shows the fan structure formed during loading. The right specimen demonstrates that during unloading all domino blocks of the fan structure rotate backwards and finally form the conventional domino-like structure that can be seen in natural faults formed in intact rocks (see photographs in **Figure 5b** [29]).

Due to the fact that direct experimental studies of the fan-structure properties are impossible, we will analyse them using a physical model. A video demonstrating the fan-structure formation and propagation along the model can be viewed at [30]. **Figure 6a** shows images of the physical model reflecting different stages of the fan-structure formation and propagation. **Figure 6a-I** shows the initial structure of the future fault that is represented by 'predetermined' domino blocks inclined at an angle α_0 to the rupture plane. All blocks made from tiles are glued together to simulate a 'monolithic' material. The bond strength is less than that of the block material. The row of domino blocks is located between two layers, AB and CD, representing the two opposite faces of the rupture (elastic connectors). The upper and lower faces are fixed to the corresponding ends of each domino block. The entire row of blocks is loaded with an evenly distributed weight, σ_p.

For the physical model, the easiest way to apply shear stress τ uniformly distributed along the entire model is the inclination of the model by an angle γ. The distributed weight, σ_p in this case, creates a shear stress $\tau = \sigma_p \sin(\gamma)$ and normal stress $\sigma_n = \sigma_p \cos(\gamma)$ along the model. By changing the angle γ, we can vary the applied stresses. Experiments conducted on the physical model show that at angle $\gamma \approx 80^0$, the upper face AB can move relative to the lower face CD due to the simultaneous separation from each other (tearing off) of all glued together domino blocks with the subsequent rotation of these blocks. We will consider this angle $\gamma \sim 80^0$ as corresponding to the material strength. At the absence of the domino structure, the frictional strength between the upper AB and lower BC faces corresponds to angle $\gamma = 40^0$.

However, if we form the fan structure from the domino blocks involved in the model, the upper face AB can be moved against the lower face CD at very low angles γ indicating very low shear resistance of the fan structure. Horizontal lines in **Figure 6b** indicate symbolically different levels of shear stresses: τ_s corresponds to the material strength, τ_f corresponds to the frictional strength, τ_{fan} corresponds to the fan-structure strength and τ corresponds to a low level of applied shear stress that can cause the fan structure to propagate along the model.

It should be emphasised that for the initial formation of the fan structure, an additional local shear stress should be applied. In the physical model, the initial fan

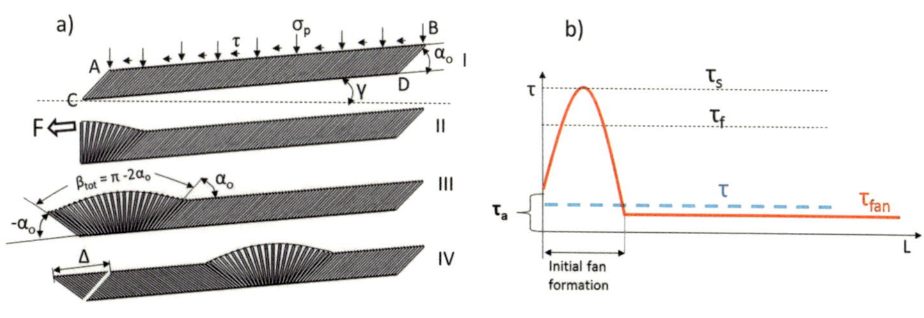

Figure 6.
Illustration of the physical model of the fan structure.

structure is generated by the application of force F to the leftmost domino block. When the local stress applied reaches a level τ_a, the front block will be torn off from the intact row, indicating the start of the tensile cracking process. The applied force is transmitted to the following blocks by elastically stretching the top rupture face (elastic connector), thereby causing the consecutive separation (tearing off) of the blocks and their rotation against the rupture faces. The fan formation is completed when the first block rotates to a total angle $\beta_{tot} = 180° - 2\alpha_0$ at a shear displacement Δ of the upper face. The red graph MM in **Figure 6b** reflects the experimentally determined variation in shear resistance of the developing fan structure during its formation and further propagation of the completed fan. The rising resistance up to τ_s is associated with the formation of the first half of the fan, while the decreasing resistance corresponds to the second half formation. The reason for such variation in shear resistance will be discussed in the next section.

Experiments on the physical model show that the minimum angle γ at which the fan propagates spontaneously along the model is about 4°. Because the frictional strength for this model is characterised by $\gamma \approx 40°$, we can conclude that shear resistance of the fan structure τ_{fan} is by the factor of ten less than the frictional strength: $\tau_{fan} \approx 0.1\tau_f$. It should be emphasised that the low shear resistance takes place within the zone of the moving fan head only. In front of the fan, the material is in an intact condition (strength τ_s). Behind the fan shear resistance is equal to friction (strength τ_f). The fact that the fan structure can propagate through the intact model (representing the 'intact' material) at very low shear stresses applied indicates that the material strength in this case is determined by the shear resistance of the fan structure. For the physical model, the fan structure decreases the material strength by the ratio $\tau_s/\tau_{fan} \approx 14$. We can suppose that for the rock specimen in **Figure 2d**, the fan mechanism decreases the material strength at the post-peak failure by the ratio $\Delta\sigma_s/\Delta\sigma_{tr} \approx 30$. The part of the post-peak curve in **Figure 3c** corresponding to the fan propagation through the intact rock specimen is represented by the horizontal line between stages 1 and 4.

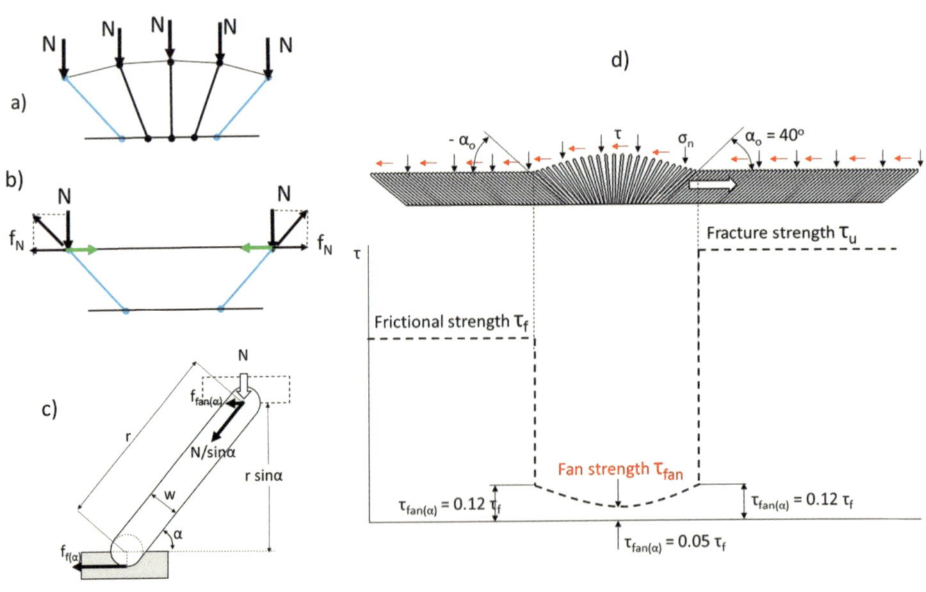

Figure 7.
Schematic explanation of the reasons for the low shear resistance of the fan structure. a) and b) Principle of self-balancing of the fan structure. c) Friction in joints of domino blocks. d) Distribution of shear resistance along the fault involving the fan structure.

In order to cause the spontaneous rupture propagation through intact rock, the fan structure should provide, in addition to low shear resistance of the rupture head, also high shear stresses in the rupture tip and sufficiently high driving power at very low shear stresses applied. The next section explains unique principles involved by the fan mechanism to satisfy these requirements.

3.3 Low shear resistance, high stress amplification and driving power generated by the fan mechanism

First, we will analyse the reason for the low shear resistance of the fan structure. Domino blocks in the fan are interconnected by the rupture faces and behave as hinges between the moving (sliding) rupture faces. Shear resistance of the fan structure τ_{fan} represents the resistance to displacement of the rupture faces relative to each other in the rupture head. **Figure 7** explains schematically the main principle responsible for low shear resistance of the fan structure. **Figure 7a** shows that all domino blocks in the fan are loaded by elementary forces N representing the normal stress σ_n. Elementary forces N applied to domino blocks in the front part of the fan resist to rotation of them. At the same time, elementary forces N applied to domino blocks in the rear part of the fan assist to rotation of these blocks. The key feature of the fan structure is the fact that each domino block in the front part of the fan is balanced by a symmetrical block of the rear part of the fan (**Figure 7b**). This means that the resistance to shear of this structure even at very high levels of normal stress applied will be determined solely by friction in joints.

Figure 7c allows estimating roughly friction in joints. It shows a self-balancing domino block (representing the right block in **Figure 7b**) of length r and width w with cylindrical ends rotating with sliding friction in corresponding cylindrical grooves. The block is inclined at an angle α and loaded by force N. An analysis conducted in [20] shows that the presence of domino blocks between the shearing rupture faces decreases the resistance to shear compared with the conventional frictional sliding in accordance with Eq. (2):

$$\tau_{fan(\alpha)} = \tau_f \, w/r \, 2 \sin^2 \alpha \tag{2}$$

Eq. (2) shows that the resistance to shear $\tau_{fan(\alpha)}$ between the rupture faces separated by self-balancing domino blocks and loaded by normal force N is determined by the conventional sliding friction τ_f, the ratio w/r and the angle α of the block inclination. Using Eq. (2) we can estimate the character of distribution of shear resistance between two rupture faces along the fan structure consisting of domino blocks characterised by the ratio w/r = 0.1 and by the initial (and final) angle $\alpha_o = 40^\circ$ as shown in **Figure 7d**. According to Eq. (2), the largest shear resistance $\tau_{fan(\alpha)} = 0.12\,\tau_f$ is provided by the front and the rear domino blocks of the fan. The minimum shear resistance $\tau_{fan(\alpha)} = 0.05\,\tau_f$ is at the middle of the fan. The dotted curve in **Figure 7d** shows the distribution of shear resistance along the fan structure. In front of the fan, shear resistance is determined by the material strength τ_s, and behind the fan it corresponds to the frictional strength τ_f. Due to a small variation in shear resistance in the fan zone, we can describe the average strength of the fan structure by Eq. (3):

$$\tau_{fan} \approx \tau_f \, w/r \tag{3}$$

For the ratio w/r = 0.1, we will have

$$\tau_{fan} \approx 0.1\tau_f \tag{4}$$

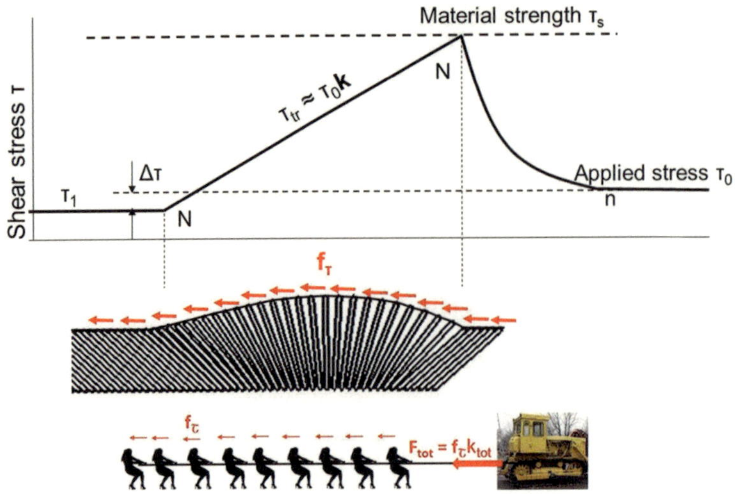

Figure 8.
Principles of shear stress amplification by the fan mechanism.

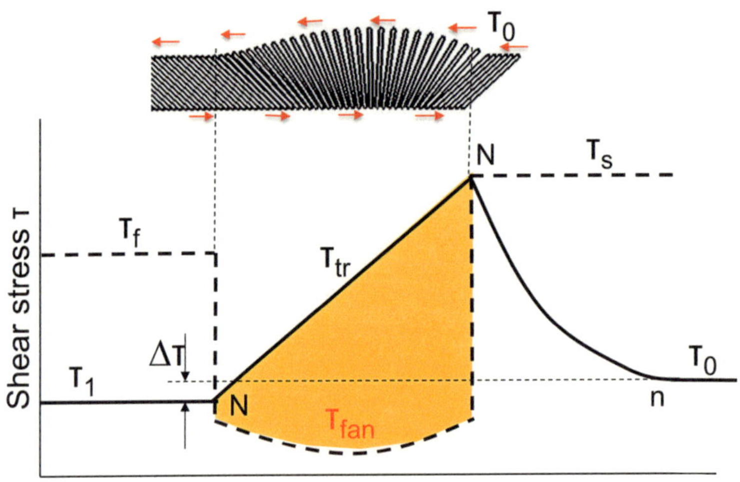

Figure 9.
Relative distribution of shear resistance and amplified shear stresses in the rupture head caused by the fan mechanism.

This estimation is in accord with experimental results obtained on the physical model.

Figure 8 illustrates schematically the unknown before and very powerful principle of shear stress amplification inherent in the fan mechanism. It shows the fan structure propagating from left to right through the intact material under the effect of distributed shear stress τ_0 applied to the whole fault. Behind the fan, domino blocks completed their rotation and form the frictional zone. The horizontal line τ_0 on the graph above the fan indicates the level of applied shear stress that is significantly less than the material strength τ_s. The explanation on how the fan mechanism can provide high stresses equal to τ_s in the rupture tip at very low shear stresses applied τ_0 is as follows.

All domino blocks of the fan structure are loaded by elementary forces f_τ representing the applied shear stress. Within the fan zone, domino blocks are separated. Due to this, elementary forces f_τ applied to each block cause corresponding rotation of them and stretch the elastic connector (i.e. the upper rupture face) in front of each block, thus transmitting these forces to the rupture tip. This principle is similar to one shown below the fan. According to this principle, real forces affecting domino blocks within the fan zone increase towards the front as shown on the graph above the fan structure (line NN). This means that shear stress within the fan zone increases from the level τ_1 at the rear end of the fan towards the rupture tip approximately in proportion to the number of domino blocks k involved in the fan structure. We will call this stress as the fan-transient stress: $\tau_{tr} \sim \tau_1 k$. Analysis conducted in [18–20] shows that shear ruptures propagating in real materials can involve the fan structure with about a thousand of domino blocks. This means that the potential ability of the fan mechanism in stress amplification can be very large, but the real generated stresses are limited by the material strength.

Figure 9 illustrates the relation between shear resistance and stresses generated by the fan mechanism at low shear stresses applied. A shaded curve here reflects the distribution of shear resistance along a fault involving the fan structure. In front of the fan, shear resistance corresponds to the material strength τ_s, behind the fan it is determined by friction τ_f and in the fan zone shear resistance is significantly less than the frictional strength $\tau_{fan} \ll \tau_f$. A solid curve reflects the shear stress variation. This schema shows that the fan mechanism can provide the rupture development at very low shear stresses applied τ_0 which can be significantly below the frictional strength, i.e. at $\tau_{fan} < \tau_0 \ll \tau_f$. The coloured area here represents symbolically the power generated by the fan mechanism that causes spontaneous rupture development even at low shear stresses applied. The instability is resulted from the fact that in the fan zone the generated stress exceeds the shear resistance and in the rupture tip it corresponds to the material strength.

It should be emphasised that despite the fact that the shear rupture here propagates through 'intact' material, the magnitude of stress drop $\Delta\tau = \tau_o - \tau_1$ can be very low (lower than at the frictional stick-slip instability) because the rupture propagates at low shear stresses applied $\tau_0 \ll \tau_f$. Depending on the level of τ_0, the fan mechanism can provide two types of rupture mode (crack-like and pulse-like) observed in natural and laboratory earthquakes [1, 31, 32]. This question

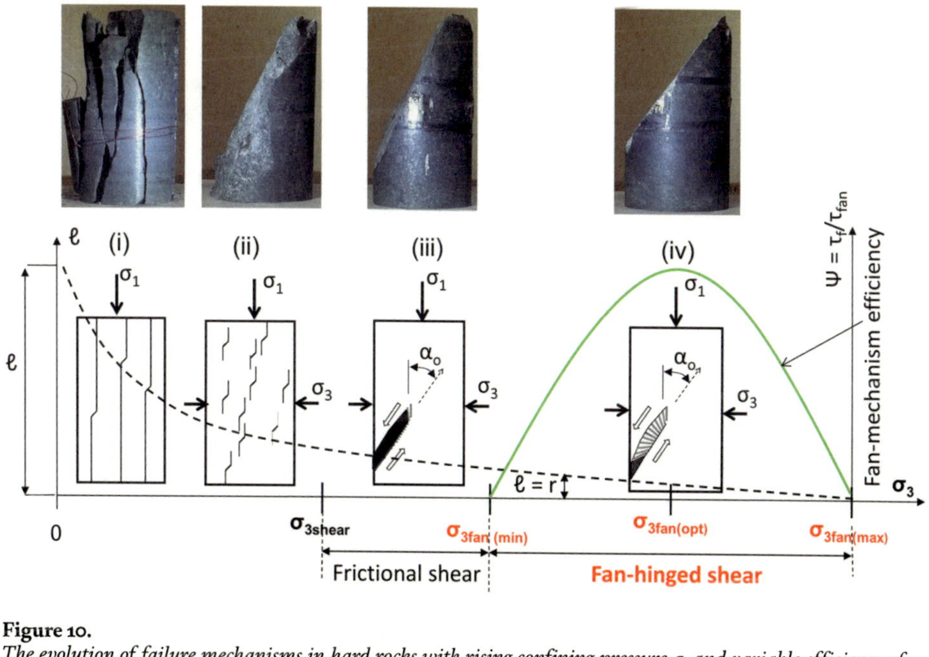

Figure 10.
The evolution of failure mechanisms in hard rocks with rising confining pressure σ_3 and variable efficiency of the fan mechanism within the pressure range $\sigma_{3fan(min)} < \sigma_3 < \sigma_{3fan(max)}$.

is discussed in [19, 20]. The fan mechanism explains also the heat flow paradox observed for extreme ruptures [6, 13].

3.4 Reasons for the fan-mechanism operation at high σ_3 and new strength profiles for hard rocks

Figure 10 illustrates the evolution of failure mechanisms in hard rock specimens with rising σ_3. Confining pressure σ_3 increases along the horizontal axis from left to right. At the origin of the horizontal axis, $\sigma_3 = 0$. The basic point is that the failure process of brittle rocks at any level of σ_3 is accompanied by formation of tensile cracks; however, the ultimate length ℓ of tensile cracks that can be developed at failure depends on the level of σ_3 because rising σ_3 suppresses the tensile crack growth. A dotted curve here shows symbolically the typical variation of ultimate length ℓ of tensile cracks as a function of σ_3: the higher the σ_3, the shorter the ℓ. The length ℓ of tensile cracks in turn determines the macroscopic failure mechanism and the failure pattern shown schematically in rock specimens (i)–(iv).

Within the pressure range $0 \leq \sigma_3 < \sigma_{3shear}$, shear rupture cannot propagate in its own plane due to creation at the rupture tip of relatively long tensile cracks that prevent the shear rupture propagation. The following two basic principles of the failure process taking place within this pressure range may be distinguished:

i. Splitting by long tensile cracks (at low σ_3)

ii. Distributed microcracking followed by coalescence of microcracks (at larger σ_3)

At $\sigma_3 \geq \sigma_{3shear}$ the failure mode is localised shear. Here high enough confining pressure σ_3 suppresses the formation of long tensile cracks, and tensile cracks generated in the rupture tip become sufficiently short to assist shear rupture to propagate in its own plane. The fracture front moves through the rock due to creation of an echelon of micro-tensile cracks in the fracture tip and inter-crack slabs (domino blocks) which are subjected to rotation at shear displacement of the rupture faces. Here, depending on behaviour of domino blocks at rotation, two basic principles of the failure process may be distinguished as discussed in **Figure 4**:

iii. Frictional shear—characterised by collapse of insufficiently short domino blocks (at $\sigma_{3shear} \leq \sigma_3 \leq \sigma_{3fan(min)}$)

iv. Fan-hinged shear—associated with formation of the fan structure consisting of sufficiently short domino blocks (of length $r = \ell$ and width w) and withstanding rotation without collapse (at $\sigma_3 > \sigma_{3fan(min)}$)

An important fact is that the fan mechanism exhibits different efficiency depending on the level of σ_3. We will determine the fan-mechanism efficiency as the ratio between the frictional strength and the fan strength: $\psi = \tau_f/\tau_{fan}$. The point is that the length r of domino blocks decreases with rising σ_3. At confining pressure near $\sigma_{3fan(min)}$, when the relative length (r/w) of domino blocks is still relatively large, the blocks are subject to partial destruction as they rotate. In this case the fan-mechanism efficiency is quite low. At higher σ_3, with shorter blocks, this imperfection decreases, rendering the fan mechanism more efficient. The optimal efficiency takes place at $\sigma_{3fan(opt)}$ when the blocks with an optimal ratio r/w rotate with minimum destruction. At greater σ_3 the efficiency reduces because shorter blocks gradually lose their potential for operation as hinges. Finally very short blocks lose this capability completely, and the rock behaviour returns to the conventional frictional mode. This happens at $\sigma_{3fan(max)}$. The green curve in **Figure 10** illustrates

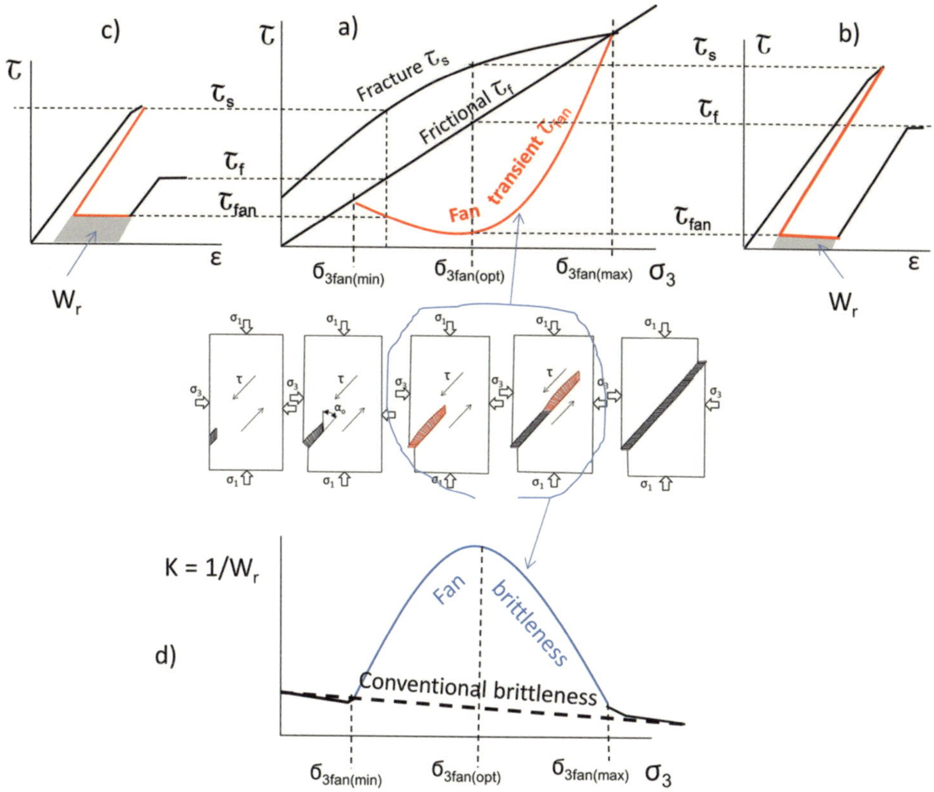

Figure 11.
Strength profiles and brittleness profiles for hard rocks. a), b) and c) Strength profiles for τ_s, τ_f and τ_{fan} plotted on the basis of complete stress-strain curves. d) Difference between the conventional and new understanding of rock brittleness variation with rising σ_3.

graphically a possible variation of the fan-mechanism efficiency $\psi = \tau_f/\tau_{fan}$ versus confining stress σ_3.

The variable efficiency of the fan mechanism with σ_3 causes corresponding variation in the transient strength and brittleness of hard rocks. **Figure 11a** shows schematically an improved model of strength profiles involving conventional fracture τ_s and frictional τ_f strength profiles and the transient fan strength τ_{fan} profile (red curve). Stress-displacement curves in **Figure 11b** and **c** reflecting real post-peak rock properties (shown in red) explain the meaning of the fan strength profile and its variation versus σ_3. **Figure 11b** corresponds to the situation at $\sigma_3 = \sigma_{3fan(opt)}$ where the fan mechanism exhibits the maximum efficiency. At peak stress the material strength is τ_s. After completion of the fan structure, the specimen strength decreases to the level τ_{fan} representing the transient material strength. This means that the shear fracture governed by the fan mechanism can propagate through the material at any level of shear stress above τ_{fan}. After the fan has crossed the specimen body, the specimen strength is determined by the frictional strength τ_f of the new fault. For confining pressures $\sigma_3 < \sigma_{3\,fan(opt)}$ or $\sigma_3 > \sigma_{3\,fan(opt)}$ the fan-mechanism efficiency is lower. The graph in **Figure 11c** indicates the relative values of τ_s, τ_f and τ_{fan} and their positions on the profiles for the situation where $\sigma_3 < \sigma_{3\,fan(opt)}$.

Let us analyse how the variable efficiency of the fan mechanism can affect the rock brittleness. In the simplest case, the brittleness of the same rock for different testing conditions can be estimated as the reciprocal of the specific rupture energy associated with the rupture propagation governed by the fan mechanism through

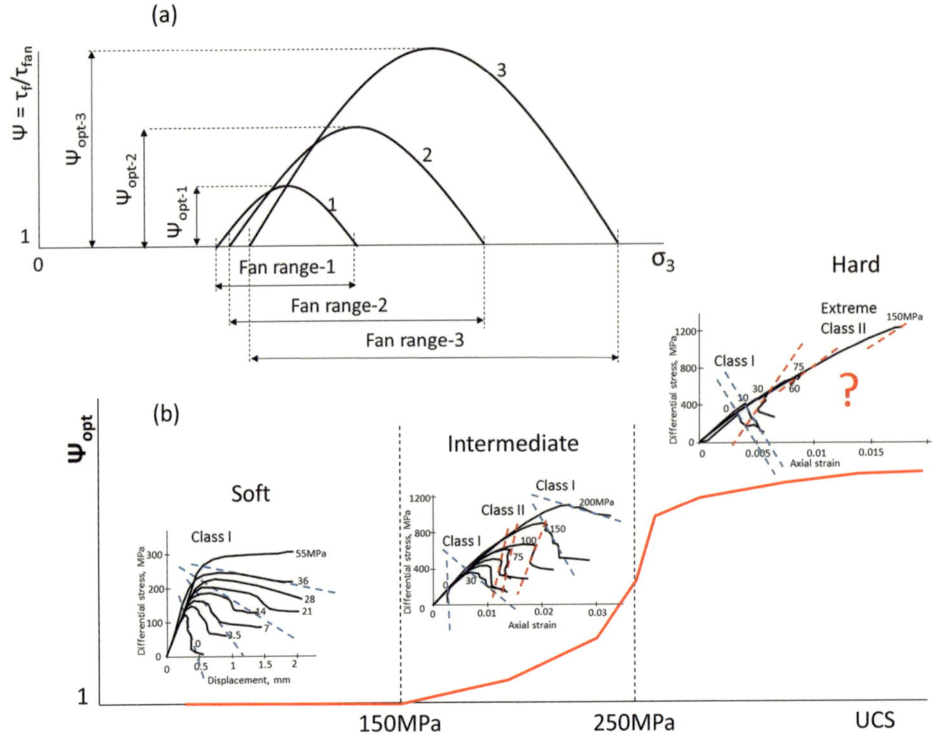

Figure 12.
(a) Variation of the fan-mechanism efficiency $\psi = \tau_f/\tau_{fan}$ versus confining pressure σ_3 for rocks of different hardness (UCS). (b) Variation of the optimal efficiency of the fan mechanism ψ_{opt} versus UCS.

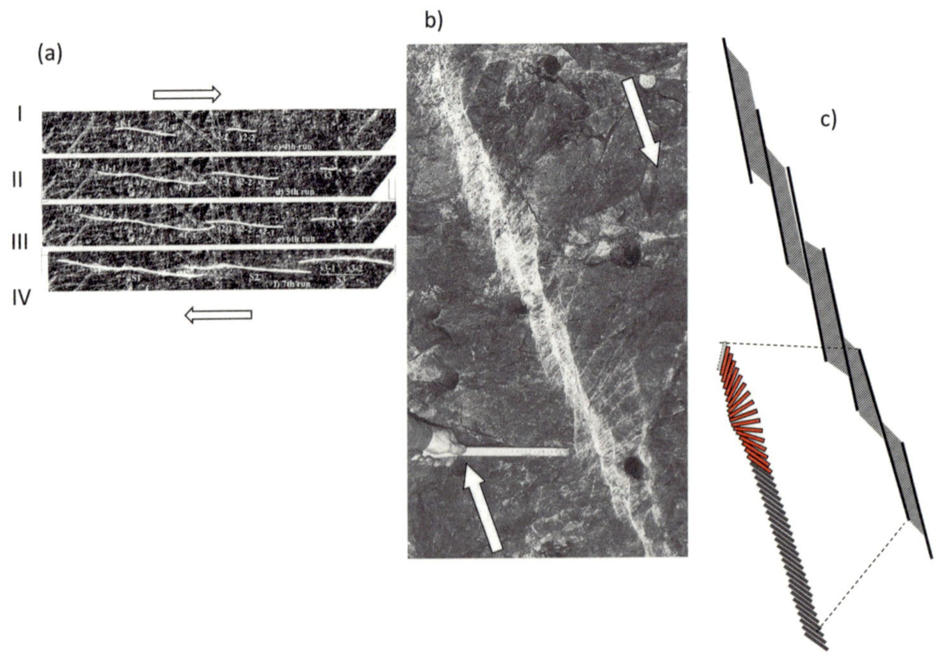

Figure 13.
a) Shear rupture propagation by advanced triggering of new segments (modified photograph from [29]). b) and c) Principle of formation of the domino and fan structure in segmented faults.

intact rock. Shaded areas on stress-strain curves in **Figure 11b** and **c** represent the specific rupture energy W_r for different σ_3. The brittleness index $K = 1/W_r$ can be used to characterise the brittleness variation versus σ_3. The shaded line in **Figure 11d** illustrates symbolically the conventional understanding of rock brittleness variation in accordance with which the rising confining pressures σ_3 make rock less brittle. Because the fan mechanism decreases dramatically the specific rupture energy within the pressure range $\sigma_{3fan(min)} < \sigma_3 < \sigma_{3fan(max)}$, the brittleness at these stress conditions should be significantly higher than the conventional understanding. The blue curve in **Figure 11d** indicates the character of rock embrittlement caused by the fan mechanism. Estimations made in [17, 22] show that at high σ_3 rock brittleness can be a hundred of times higher than at low σ_3. At $\sigma_3 = \sigma_{3fan(opt)}$ rock conditions can be super brittle.

The fan-mechanism efficiency $\psi = \tau_f/\tau_{fan}$ depends also on the rock hardness characterised by UCS. **Figure 12a** shows three curves indicating variations of $\psi = \tau_f/\tau_{fan}$ versus σ_3 for three rocks of different hardness. Here the harder the rock, the greater the optimal fan-mechanism efficiency ψ_{opt} and the larger the range of σ_3 where the fan mechanism is active. The red curve in **Figure 12b** illustrates symbolically the dependence of the optimal efficiency of the fan mechanism ψ_{opt} on the hardness (UCS) of different rocks. The fan mechanism operates with the largest efficiency in rocks with UCS > 250 MPa. Within the range of UCS 150–250 (roughly), the efficiency is significantly lower. In soft rocks the fan mechanism is not active.

4. Fan mechanism as a source of dynamic events in the earth's crust

4.1 Features of the fan-structure formation in complex natural faults

This section discusses the role of the fan mechanism in generation of shallow earthquakes and shear rupture rockbursts in deep mines. In the previous sections, we introduced different unique features of the fan mechanism and 'abnormal' properties of hard rocks. All analysis was conducted for primary shear ruptures which are thin and continuous. Unlike primary ruptures natural faults typically have very complicated segmented and multi-hierarchical structure [7, 33, 34]. Main principles of the complex fault evolution in association with the fan mechanism were discussed in [20, 23, 35]. Here we will outline briefly most important features of the fan-mechanism generation in complex faults.

It was observed that in ultra-deep South African mines, very severe dynamic events (shear rupture rockbursts) are caused by new shear ruptures generated in pristine rock [36, 37]. These mine tremors are seismically indistinguishable from natural earthquakes and share the apparent paradox of failure under low shear stress [37]. Photographs of such faults are shown in **Figure 5b**. The structure of all these faults is identical consisting of a row of domino blocks. However, the domino structure is more complex than in primary ruptures.

Figure 13 explains features of this structure formation. Series of photographs in **Figure 13a** (modified from [38]) shows principles of segmented fault propagation observed experimentally. The fault propagates due to advanced triggering of new segments. The photographs show four stages (I–IV) of the fault evolution. Segments are represented here by white lines. The fault propagates from left to right. The segments are generated one by one due to the stress transfer and propagate bilaterally. At the meeting of each two neighbouring segments, they are connected by a compressive jog. It was found out in [29, 38] that jogs of the compression type are very common at high confining pressures to fault zones regardless of their sizes. Overlap zones of the jogs

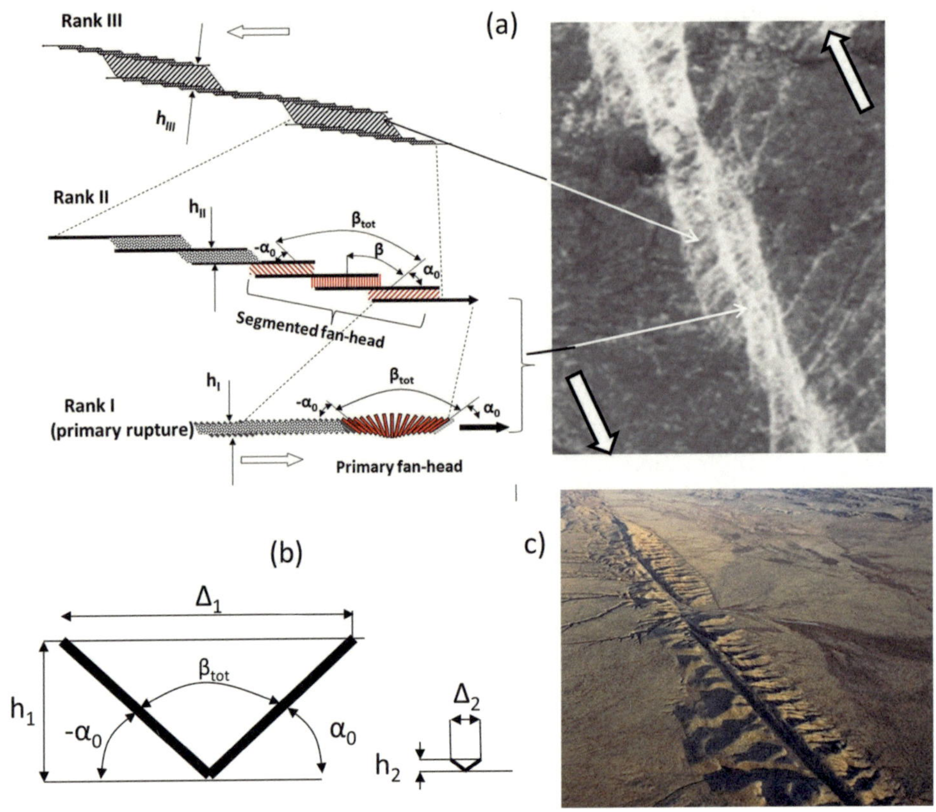

Figure 14.
The fan mechanism is predominantly activated in fault segments of lower hierarchical ranks (schematic illustration) providing extreme dynamics along thin localised zones.

are subjected to significant irreversible deformation. **Figure 13b** and **c** demonstrates that in brittle rocks the irreversible deformation in jogs is associated with formation of a row of domino blocks (modified photograph from [29]). The general fault here consists of a number of segments represented by primary ruptures. The propagation of primary ruptures in hard rocks at high σ_3 is governed by the fan mechanism.

The domino structure of the next hierarchical ranks can also be involved in the fan-structure formation. This feature is illustrated in **Figure 14**. It was observed in [38] that segmentation as a mechanism of fault propagation acts on all hierarchical ranks of complex faults. Once a number of segments of a given hierarchical rank coalesce, they behave as a whole as a new and longer segment of one higher rank. Segment of higher rank can trigger a new segment (shear fracture) at greater distance. A photograph in **Figure 14a** (modified from [29]) shows a fault fragment involving segments of three hierarchical ranks. The structure of this fault is shown symbolically on the left. It incorporates primary ruptures and higher rank segments formed on the basis of compressive jogs represented by the domino structure (rank II and rank III). Domino blocks involved in segments of higher rank can form the fan structure similar to primary ruptures due to rotation of them caused by shear displacement of the rupture faces. However, the complete fan structure can be formed if shear displacement between the fault faces d_{fault} is sufficient for the completed block rotation.

Figure 14b shows the initial and final positions of domino blocks for two shear ruptures of thicknesses h_1 and h_2. The thick rupture requires significantly greater displacement $d_{fault} = \Delta$ to complete the block rotation. Due to this, the complete fan structure (red zones in **Figure 14a**) is predominantly created in segments of lower ranks. The fan mechanism generated here creates high dynamics of the failure process. Relatively thin localised zones of very intense destruction can be observed in each dynamic fault. The initial domino structure of these segments is completely destroyed by extensive and violent shear and represented by pulverised gouge. In high-rank segments, domino blocks rotate by low angles without formation of the fan structure and assist the accommodation of displacement along the whole fault. The domino-like structure is typical for faults of very different scales including laboratory specimens, shear rupture rockbursts in mines and earthquakes. Two photographs in **Figure 14a** and **c** show identical domino structure of two dynamic faults which generated severe shear rupture rockburst (a South African ultra-deep mine) and earthquake (the San Andreas fault exposed on land) [39].

4.2 Generation of new faults in intact rock at low shear stresses nearby a pre-existing fault caused by the fan mechanism

This section proposes an alternative explanation to the fact that earthquakes are commonly attributed to pre-existing faults. Pre-existing discontinuities play the role of local stress concentrators, creating the starting conditions for the fan-structure formation. After completion of the initial fan structure, it can create a new dynamic fault in the form of earthquake by propagation through intact rock mass loaded by low shear stresses. **Figure 15** illustrates one of the many models for generation of high local stress on the basis of pre-existing fault. It shows a rock fragment involving a pre-existing fault (black line) with a compressive jog. This

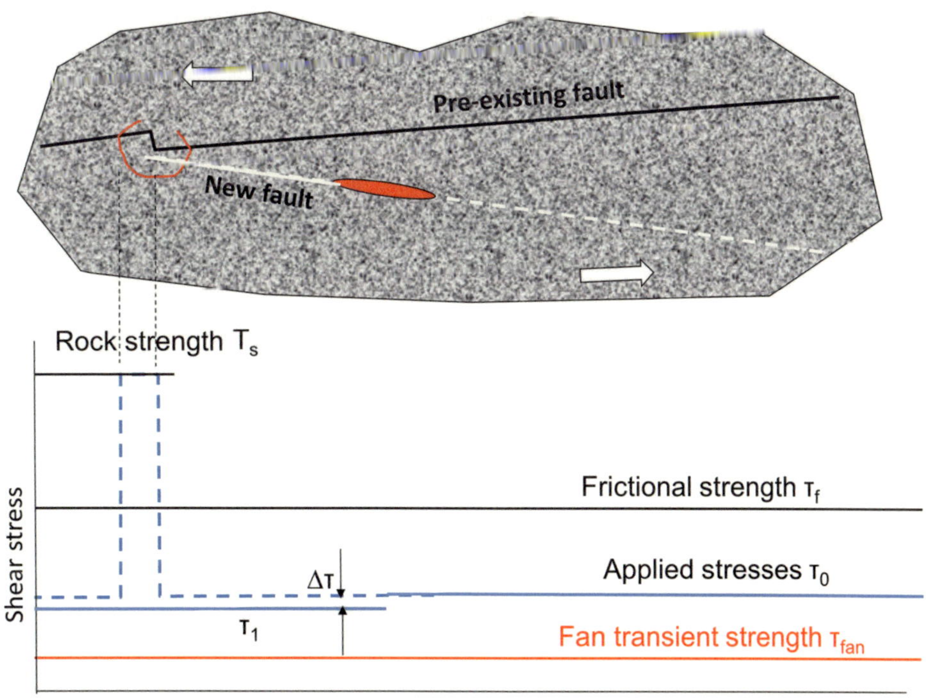

Figure 15.
Features of generation of a new extreme rupture in pristine hard rock in the vicinity of a pre-existing fault at low field shear stresses caused by the fan mechanism.

fragment is located at great depth where the minor stress σ_3 is high enough for the fan-mechanism activation in intact rock. Horizontal lines on the graph below indicate symbolically levels of the following parameters: τ_s is the strength of intact rock, τ_f is the frictional strength of pre-existing fault, τ_{fan} is the transient strength of intact rock determined by the fan mechanism, τ_0 is the field shear stress applied to the rock fragment, τ_1 is the field stress after the rupture propagation and $\Delta\tau$ is the stress drop. Orientation of the field shear stress is shown by open arrows.

Because the level of field stress τ_0 is significantly less than the frictional strength τ_f, the situation on the pre-existing fault is very stable. However, due to deformations along the fault caused by shear stresses τ_0, a high local stress can be created in the jog zone delineated by a red circle. If the local stress in intact rock of this zone reaches the level of rupture strength τ_s, the fan structure can be formed. After formation of the fan structure, it can propagate spontaneously through intact rock mass at low shear stresses τ_0 in accordance with Class III behaviour discussed in **Figure 3** and generate an earthquake. The new fault is shown by a white line, and the propagating fan head is represented by red ellipsis. Due to very high brittleness of this rock associated with extremely low rupture energy provided by the fan mechanism, the failure process can be accompanied by abnormal energy release and violence. It should be emphasised that despite the fact that the new fault is formed in intact rock, the magnitude of stress drop $\Delta\tau$ can be very low because this process takes place at low shear stress applied. The stress drop can be even less than at the stick-slip process in the case of activation of the pre-existing fault.

Thus, the fan mechanism favours the generation of new faults in hard intact rock mass adjoining a pre-existing fault in preference to frictional stick-slip instability along the pre-existing fault. Each earthquake generated by the fan mechanism is associated with formation of a new fault at a new location in the vicinity of a pre-existing fault. Furthermore, each new fault can serve as a stress concentrator for generation of the next new fault. The proximity of the pre-existing fault to the zone of dynamic new fracture development in intact rock creates the illusion of frictional stick-slip instability of the pre-existing fault, thus concealing the real situation.

At the same time, there are many evidences that earthquakes are associated with the formation of new faults in the proximity of pre-existing faults. For example, **Figure 16** shows maps of earthquakes in a New Zealand region of relative motions between the Australian and Pacific plates which are not accommodated on one general fault, but on many faults across a wide zone. **Figure 16a** (from [40]) shows

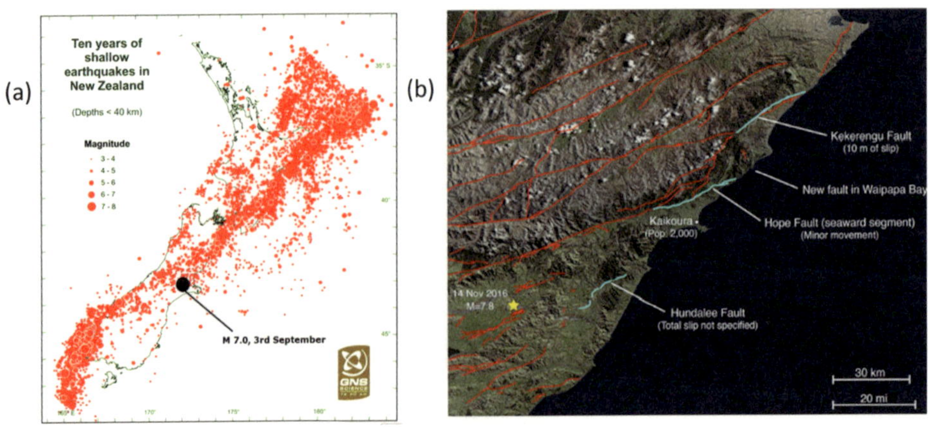

Figure 16.
Maps of spatial distribution of earthquake hypocentres and faults on the earth's surface for a New Zealand region [40, 41].

spatial distribution of hypocentres at depths <40 km. Some earthquake faults generated at great depths can reach the earth's surface as shown in **Figure 16b** (from [41]). The formation of each fault from reaching the earth's surface is associated with an earthquake. The fan mechanism which makes intact hard rocks weaker in respect of dynamic shear rupturing than pre-existing faults is responsible for the spatial distribution of earthquake hypocentres and for the fact that the earth's crust is riddled with faults. A series of aftershocks, which usually accompany the main act of an earthquake, can also be explained by the formation of a series of new faults, where each new fault creates the conditions for activating the fan mechanism in the adjacent zone of intact rock.

4.3 Depth distribution of rock strength, brittleness and earthquake activity caused by the fan mechanism

Figure 17d shows a typical histogram of depth distribution of earthquake frequency (from [7]). It demonstrates that earthquake activity varies with depth and has a maximum at a certain depth. Today there are two fundamentally different explanations for this earthquake feature. Both of them consider earthquakes as stick-slip instability on pre-existing faults. The first one is based on the fact that the frictional strength (determining the lithospheric strength) in the upper crust increases with depth in accordance with Byerlee's friction law [10], while in the lower crust it decreases accordingly to a high-temperature steady-state flow law [2, 7]. The second one is based on the velocity-weakening and velocity-strengthening concept [7, 9]. We introduce here a new concept which is based on the new understanding about (unknown before) properties of hard rocks at seismic depth's caused by the fan mechanism [20, 23, 42].

Figure 17a–c shows symbolically depth distribution of the fan-mechanism efficiency, rock strength profiles and rock brittleness. These graphs are analogous to the dependencies discussed in **Figures 11** and **12**. The fan mechanism can operate at depths where temperature (rising with depths) does not prevent the fan-hinged shear. The new strength profile for hard rocks in **Figure 17b** shows that at low depths corresponding to $\sigma_3 < \sigma_{3fan(min)}$, the lithospheric strength is determined solely by frictional strength τ_f. At greater depths corresponding to the range of the fan-mechanism activity, the situation is specific. In the absence of conditions for

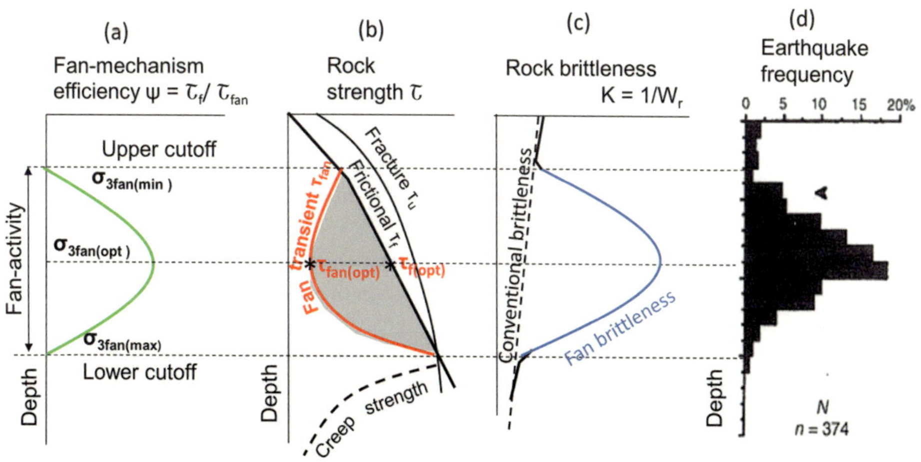

Figure 17.
Relation between depth distribution of the fan-mechanism efficiency, rock strength, rock brittleness and earthquake frequency.

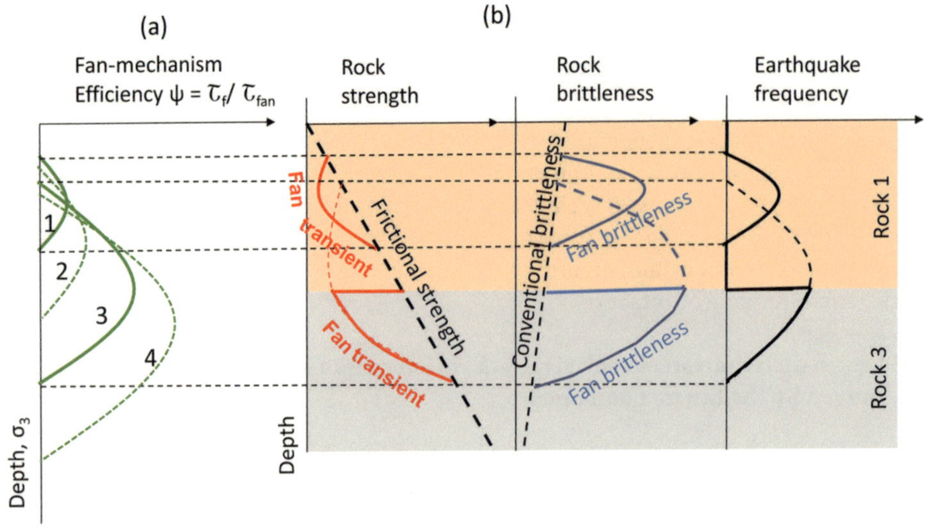

Figure 18.
Illustration of depth distribution for rock transient strength, brittleness and earthquake frequency in the Earth's crust represented by two layers of rocks of different hardness.

activation of the fan mechanism, the lithospheric strength is determined by friction. However, if the fan mechanism is activated somewhere causing the new rupture development in intact rock, the transient lithospheric strength in that region decreases to the level τ_{fan}. After completion of the failure process, the lithospheric strength returns to the frictional strength.

It should be noted that the improved concept of the lithospheric strength incorporates all three types of rock strength determining the instability in the seismic layer: fracture strength τ_s, frictional strength τ_f and fan strength τ_{fan}. The fracture strength determines the level of local stress at which the initial fan structure can be generated. The shaded area between the frictional τ_f and the fan-transient τ_{fan} profiles determines levels of field stress under which an initiated fan structure can propagate creating an earthquake. Importantly, the fan mechanism can cause earthquakes at any level of field stress τ within the shaded zone. Due to this the highest probability of events is at a depth characterised by the maximum range between τ_{fan} and τ_f. This depth corresponds approximately to the depth of optimal efficiency of the fan mechanism, where the rock mass is characterised by the minimum transient strength and maximum brittleness. At lower and greater depths, the probability decreases. This feature determines the typical depth-frequency distribution of earthquake hypocentres. The upper and lower cut-offs represent boundaries of the zone of the fan-mechanism activity. The explanation for the depth distribution of earthquake frequency on the basis of the fan mechanism differs fundamentally from the conventional explanations.

On the basis of the fan mechanism, it is possible also to explain the existence of a few zones of earthquake activity with depth. As discussed in **Figure 12**, the efficiency of the fan mechanism depends on the rock hardness (UCS): the harder the rock, the greater the fan-mechanism efficiency and the wider the confining pressure range over which the fan mechanism is active. **Figure 18a** illustrates schematically depth distributions of the fan-mechanism activity for four rocks characterised by different hardness, with strength increasing from rock 1 to rock 4. **Figure 18b** shows a situation when the earth's crust is represented by two layers of rocks of different hardness (rock 1 and rock 3). In this case, two zones of earthquake activity may be observed. Rock 1 will exhibit the typical (complete) form of earthquake frequency-depth distribution, while rock 3 will show a truncated form. Such features of earthquake behaviour have been observed in nature and explained on the basis of conventional velocity-weakening and velocity-strengthening approach [9].

The fan theory proposes new explanations for a number of other abnormalities and paradoxes associated with extreme rupture dynamics (including supershears) observed in natural and laboratory conditions [19, 20, 43]. Mathematical models were developed which allow studying unique features of the fan mechanism and simulating the process of extreme rupture development at different loading conditions [43–46].

Acknowledgements

This work was supported by the Ministry of Science and Education of the Russian Federation (grant no. RFMEFI58418X0034).

Author details

Boris Tarasov
The Far Eastern Federal University, Vladivostok, Russia

*Address all correspondence to: bgtaras@gmail.com

References

[1] Brace WF, Byerlee JD. Stick-slip as a mechanism for earthquakes. Science. 1966;**153**(3739):990-992

[2] Brace WF, Kohlstedt D. Limits on lithospheric stress imposed by laboratory experiments. Journal of Geophysical Research. 1980;**85**:6248-6252

[3] Ruina AL. Slip instability and state variable friction laws. Journal of Geophysical Research. 1983;**88**:10359-10370

[4] Heaton TH. Evidence for and implications of self-healing pulses of slip in earthquake rupture. Physics of the Earth and Planetary Interiors. 1990;**64**(1):1-20

[5] Dieterich JH. Earthquake nucleation on faults with rate-dependent and state-dependent strength. Tectonophysics. 1992;**211**:115-134

[6] Ben-Zion Y. Dynamic ruptures in recent models of earthquake faults. Journal of the Mechanics and Physics of Solids. 2001;**49**:2209-2244

[7] Scholz CH. The Mechanics of Earthquakes and Faulting. Cambridge: Cambridge University Press; 2002

[8] Allmann BP, Shearer PM. Global variations of stress drop for moderate to large earthquakes. Journal of Geophysical Research. 2009;**114**:B01310. DOI: 10.1029/2008JB005821

[9] Albaric J, Deverchere J, Petit C, Perrot J, Le Gall B. Crustal rheology and depth distribution of earthquakes: Insights from the central and southern East African Rift System. Tectonophysics. 2009;**468**:28-41

[10] Byerlee JD. Friction of rocks. Pure and Applied Geophysics. 1978;**116**:615-626

[11] Lachenbruch AH. Frictional heating, fluid pressure, and the resistance to fault motion. Journal of Geophysical Research. 1980;**85**:6097-6112

[12] Dieterich JH, Conrad G. Effect of humidity on time- and velocity-dependent friction in rocks. Journal of Geophysical Research. 1984;**89**:4196-4202. DOI: 10.1029/JB089iB06p04196

[13] Brown R. Frictional heating on faults: Stable sliding versus stick slip. Journal of Geophysical Research. 1998;**103**(B4):7413-7420

[14] Di Toro G, Goldsby DL, Tullis TE. Friction falls towards zero in quartz rock as slip velocity approaches seismic rates. Nature. 2004;**427**:436-439

[15] Ben-David O, Rubinstein SM, Fineberg J. Slip-stick and the evolution of frictional strength. Nature. 2010;**463**:76-79

[16] Tarasov BG. Intersonic shear rupture mechanism. International Journal of Rock Mechanics and Mining Science. 2008;**45**(6):914-928

[17] Tarasov BG. Super brittleness of rocks at high confining pressure. In: Van Sint Jan M, Potvin Y, editors. Keynote addresses on Proceedings of the Fifth International Seminar on Deep and High Stress Mining. Perth: Australian Centre for Geomechanics; 2010. pp. 119-133

[18] Tarasov BG. Hitherto unknown shear rupture mechanism as a source of instability in intact hard rocks at highly confined compression. Tectonophysics. 2014;**621**:69-84

[19] Tarasov BG. Shear fractures of extreme dynamics. Rock Mechanics and Rock Engineering. 2016;**49**(10):3999-4021

[20] Tarasov BG. Shear ruptures of extreme dynamics in laboratory and

natural conditions. In: Wesseloo J, editor. Keynote addresses on Eighth International Conference on Deep and High Stress Mining. Perth, ISBN: 978-0-9924810-6-3; 2017. pp. 1-48

[21] Tarasov BG, Randolph MF. Frictionless shear at great depth and other paradoxes of hard rocks. International Journal of Rock Mechanics and Mining Science. 2008;**45**(3):316-328

[22] Tarasov BG, Randolph MF. Super brittleness of rocks and earthquake activity. International Journal of Rock Mechanics and Mining Science. 2011;**48**:888-898

[23] Tarasov BG, Randolph MF. Improved concept of lithospheric strength and earthquake activity at shallow depths based upon the fan-head dynamic shear rupture mechanism. Tectonophysics. 2016;**667**:124-143

[24] Rummel F, Fairhurst C. Determination of the post-failure behavior of brittle rock using a servo-controlled testing machine. Rock Mechanics and Rock Engineering. 1970;2(4):189-204

[25] Reches Z, Lockner DA. Nucleation and growth of faults in brittle rocks. Journal of Geophysical Research. 1994;**99**:18159-18173

[26] Peng S, Johnson AM. Crack growth and faulting in cylindrical specimens of Chelmsford granite. International Journal of Rock Mechanics and Mining Sciences. 1972;**9**:37-86

[27] Horii H, Nemat-Nasser S. Compression-induced micro-crack growth in brittle solids: Axial splitting and shear failure. Journal of Geophysical Research. 1985;**90**:3105-3125

[28] King GCP, Sammis CG. The mechanisms of finite brittle strain. Pure and Applied Geophysics. 1992;**138**:611-640

[29] Ortlepp WD. Rock Fracture and Rockbursts. Johannesburg: The South African Institute of Mining and Metallurgy; 1997

[30] Tarasov B. Fan-hinged shear, online video, 19 July, viewed 6 December 2016. 2016. Available from: https://www.youtube.com/watch?v=_-AUzCEw35M&feature=youtu.be

[31] Lu X, Lapusta N, Rosakis AJ. Pulse-like and crack-like ruptures in experiments mimicking crustal earthquakes. Proceedings of the National Academy of Science USA. 2007;**104**:18931-18936

[32] Lu X, Rosakis AJ, Lapusta N. Rupture modes in laboratory earthquakes: Effect of fault prestress and nucleation condition. Journal of Geophysical Research. 2010, 2010;**115**:B12302. DOI: 10.1029/2009JB006833

[33] Segal P, Pollard DD. Mechanics of discontinuous faulting. Journal of Geophysical Research. 1980;**85**:4337-4350

[34] Sibson RH. Fault zone models, heat flow, and the depth distribution of earthquakes in the continental crust of the United States. Bulletin of the Seismological Society of America. 1982;**72**:151-163

[35] Tarasov BG. Fan-structure shear rupture mechanism as a source of shear rupture rockbursts. Journal of the Southern African Institute of Mining and Metallurgy. 2014;**114**(10):773-784

[36] Gay NC, Ortlepp WD. Anatomy of a mining-induced fault zone. Geological Society of America Bulletin. 1979;**90**:47-58

[37] McGarr A, Pollard D, Gay NC, Ortlepp WD. Observations and analysis of structures in exhumed mine-induced faults. U.S. Geological Survey Open File Report. 1979;**79-1**(239):101-120

[38] Otsuki K, Dilov T. Evolution of hierarchical self-similar geometry of experimental fault zones: Implications for seismic nucleation and earthquake size. Journal of Geophysical Research. 2005;**110**. DOI: B03303, 10.1029/204JB003359

[39] http://www.geologyin.com/2018/06/slow-earthquakes-on-san-andreas-fault.html

[40] Chris R. Tectonics of the M7 earthquake near Christchurch. New Zealand; 2010. Available from: http://all-geo.org/highlyallochthonous/2010/09/tectonics-of-the-m7-earthquake-near-christchurch-new-zealand/

[41] Temblor. Temblor. 2017. Available from: http://www.temblor.net

[42] Tarasov BG. Depth distribution of lithospheric strength determined by the self-unbalancing shear rupture mechanism. In: Proceedings of the International Symposium, Rock Mechanics for Resources, Energy and Environment (Eurock), Wroclaw, Poland. 2013. pp. 165-170

[43] Tarasov BG, Guzev MA, Sadovskii VM, Cassidy MJ. Modelling the mechanical structure of extreme shear ruptures with friction approaching zero generated in brittle materials. International Journal of Fracture. 2017. DOI: 10.1007/s10704-017-0223-1

[44] Tarasov BG, Guzev MA. New insight into the nature of size dependence and the lower limit of rock strength. In: Malovichko AD, editor. Proceedings 8th International Symposium on Rockbursts and Seismicity in Mines. Russia: St-Petersburg; 2013. pp. 31-40

[45] Tarasov BG, Guzev MA. Mathematical model of fan-head shear rupture mechanism. Key Engineering Materials. 2014;**592-593**:121-124. DOI: 10.4028/www.scientific.net/KEM.592-593.121

[46] Tarasov BG, Sadovskii VM, Sadovskaya OB. Analizis of fan waves in a laboratory model simulating the propagation of shear ruptures in rocks. Computational Mechanics of Solids. 2016;**9**(1):38-51. DOI: 10.7242/1999-6691/2016.9.1.4

Chapter 6

A Review of Chinese Ice Caves

Shao-Hua Yang and Yao-Lin Shi

Abstract

Ice caves are a rare geological phenomenon. Ningwu ice cave, Zibaishan ice cave, and Wudalianchi ice cave are the most famous ice caves in China. We described each one in detail and carried out thermal-elastic modeling and heat conduction modeling to investigate the stability of the cave and the formation of ice deposit. In order to quantitatively study the mechanism of formation and preservation of the ice cave, we applied the FEM to model the heat exchange in the ice cave. The modeling results revealed that there is the seasonal asymmetric energy exchange. Heat energy is conducted inefficiently into the ice cave from outside and wall rock in spring, summer, and autumn. While in winter, heat energy is transferred very efficiently due to the air natural convection, thus cooling it down. We proposed that Ningwu ice cave and Zibaishan ice cave may be a self-regulating system, respectively. At Wudalianchi ice cave, airtight doors have been installed at these ice caves' entrances. This actually prevents cooling in winter. We expect that no airtight door will be fixed at each ice cave's entrance, and few people enter the ice cave before comprehensive and detailed studies, avoiding further affecting its natural conditions.

Keywords: ice cave, air convection, energy exchange, water-ice phase change, numerical modeling

1. Introduction

Many caves developed in karstic regions and basaltic regions. But ice caves, permanent ice deposit preserved within caves, are a rare geological phenomenon. An ice cave is a type of natural cave that contains significant amounts of perennial ice. The most famous of ice caves are Eisriesenwelt ice cave, Austria [1–3]; Dobšinská ice cave, Slovakia [4, 5]; Scarisoara ice cave, Romania [6, 7]; and Monlesi ice cave, Switzerland [8, 9]. The Eisriesenwelt ice cave is the largest ice cave in the world. More than 10 ice caves have been found in China, and Ningwu, Zibaishan, and Wudalianchi ice caves are the most famous ice caves.

As long ago as 1861, studies of ice caves began [10]. In the background of global climate change, seven international conferences on ice caves have been held in recent decades [11]. Several articles have documented seasonal air temperature fluctuations of several degrees from cave systems [12–14]. Consequently, in order to approximate the impact of climatic conditions on cave environments, a better interpretation of subsurface heat transfers is necessary [9]. In addition to this, ice caves are tourism resources. A better interpretation of subsurface heat transfers may assist in managing ice caves more scientifically.

Empirical calibrations have been carried out previously to evaluate the spatio-temporal distribution of cave temperature as a mathematical relation of the outside atmospheric conditions [15, 16]. In temperate karstic natural conditions, interpretation of the existence of subsurface ice deposits represents probably the most severe test for models of the magnitude and direction of heat and mass transfers induced by cave air circulation [9]. In theory and application (mathematics and engineering), the finite element method (FEM) and the finite difference method (FDM) are classical methods for finding approximate solutions of partial differential equations, which governed physical processes (including heat transfer). The details of FEM and FDM can be found in many textbooks [17, 18]. These methods would be useful for ice cave studies.

Since Ningwu ice cave was found, ice cave studies began in 1998 in China. Ningwu ice cave in Shanxi Province has been broadly reported during the past decade [19, 20], but little was known about the physical processes controlling the formation and preservation of permanent subsurface ice deposits under temperate climate conditions [21]. In addition to, basaltic cave stability is significant for ice preservation. FEM was applied to investigate the energy exchange of Ningwu ice cave and then quantitatively interpret the formation and preservation mechanism of the ice deposit. Thermal-elastic modeling was conducted to study lava tube stability during its cooling process.

2. Geological setting and characteristics of three ice caves in China

2.1 Ningwu ice cave

Ningwu ice cave (38°57′N, 112°10′E; 2121 m above sea level (asl)) is called "ten thousand years ice cave" by local people. It is located in the shaded slope of Guancen Mountain, Ningwu County, and Shanxi Province. The host rock consists of Ordovician Majiagou limestone, dolomitic limestone, argillaceous dolomite, and thin brecciated limestone and is densely fractured [22].

As a part of Ningwu National Geological Park, Ningwu ice cave is an important tourist attraction. Above 1500 visitors enter the cave per day from May to October. It has only a single entrance. People can walk into the inside of the cave by wooden spiral stairs. Ice covers the host rock almost completely. Ice stalactites and ice stalagmites can be found in all parts of the cave (**Figure 1a**).

In order to investigate the fine geometry of Ningwu ice cave, a classic geophysical exploration (using magnetotelluric measurement) was carried out and produced a high-resolution two-dimensional vertical cross-section of the ice cave [22] (**Figure 1b**). **Figure 1b** illustrates a bowling pin-shaped room, and the cave space is about 85 m depth. The widest part is in the middle of the cave with a width of 20 m.

The outside of Ningwu ice cave keeps a temperate climate. From June to September, the mean air temperature is about 14.6°C, and the average annual air temperature is 2.3°C [20]. The nearest meteorological station to Ningwu ice cave is Wuzhai station (about 320 m lower than Ningwu ice cave), at which the daily temperature was measured continuously from 1957 to 2008. We averaged the observational air temperature at Wuzhai meteorological station to get the annual temperature and then derived the mean annual temperature at Wuzhai station. The difference between the average annual air temperature at Ningwu ice cave and that at Wuzhai station can be calculated. Finally, we obtained the annual temperature variation outside Ningwu ice cave through reducing the annual temperature at Wuzhai station by the difference.

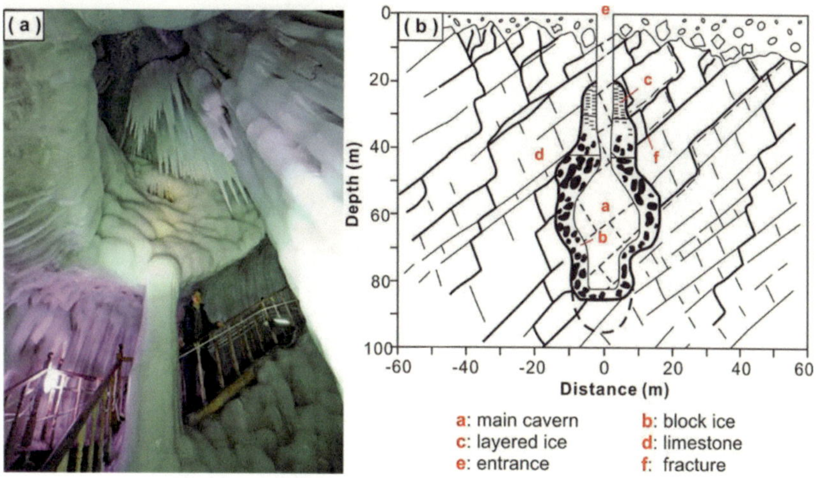

Figure 1.
2D geometry of Ningwu ice cave based on geophysical exploration [22] (a) and its interior (b).

2.2 Zibaishan ice cave

Zibaishan ice cave (33.7°N, 106.68°E; 2450 m asl) was found in August 2015, and it is located in the shaded slope of Zibai Mountain, Liuba County, Shanxi Province. Fieldwork was conducted in May 2016, and then the geometry and temperature profile of the ice cave was measured. The measurements are shown in **Figure 2a**. The cave space has a depth of about 90 m. The widest part is at the bottom with a width above 25 m.

Zibaishan ice cave, located in Zibaishan National Forest Park, is also a major tourist attraction. Most of the tourists only visited the opening of the ice cave, and a few experienced adventurers with professional climbing equipment could reach the interior of the ice cave. Consequently, Zibaishan ice cave maintained better natural environments (because of little artificial effects) than Ningwu ice cave. Ice deposits of the ice cave mainly included two parts: (1) ice cone which has a diameter of 25 m and a height of 16 m (**Figure 2b**) and (2) ice stalactites (**Figure 2c**). Ice deposits covered wall rocks of the ice cave partially. The host rock consists of limestone.

The outside of Zibaishan ice cave belongs to the warm temperate zone and humid monsoon section. The external annual mean air temperature is 11.5°C, and the frost-free period is 214 days per year [23]. Generally, ice deposits could not be preserved under such a temperate environment. Therefore, there must be some special cooling mechanisms of Zibaishan ice cave.

2.3 Wudalianchi ice cave

There are two ice caves at Wudalianchi National Geological Park (48.647°, 126.25°, 400 m asl), Heilongjiang Province: one is Bailong (means white dragon; also named Dixiabinghe) ice cave, and another is Shuijinggong (means crystal palace) ice cave.

We conducted simple field work on Bailong ice cave in August 2012. The host rock is basalt. Dense fractures developed near the entrance. Basaltic pillars can be seen within the cave. The farthest distance we can reach is about 270 m from

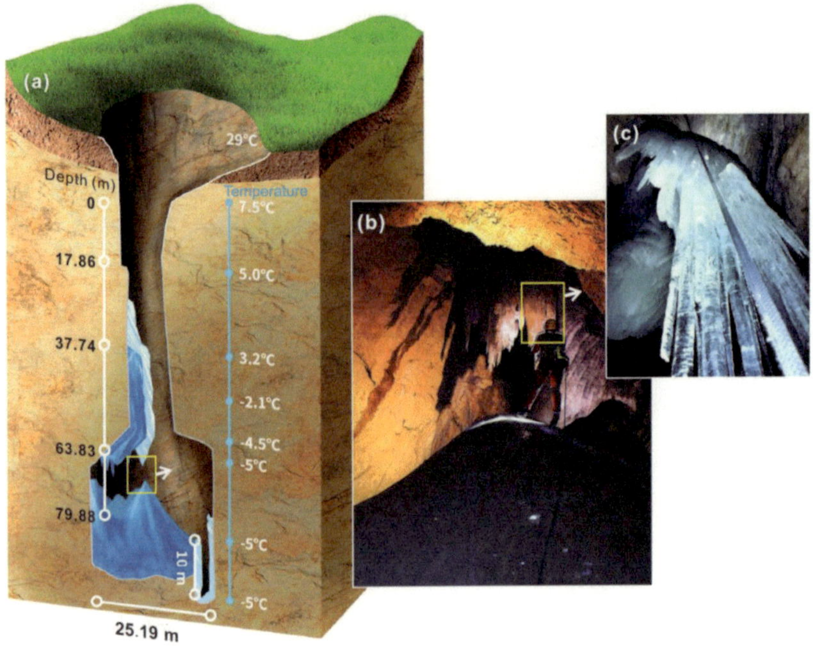

Figure 2.
The space and interior of Zibaishan ice cave (modified from [23]). (a) The geometry and temperature profile, (b) ice cone and ice stalactites, (c) ice stalactites.

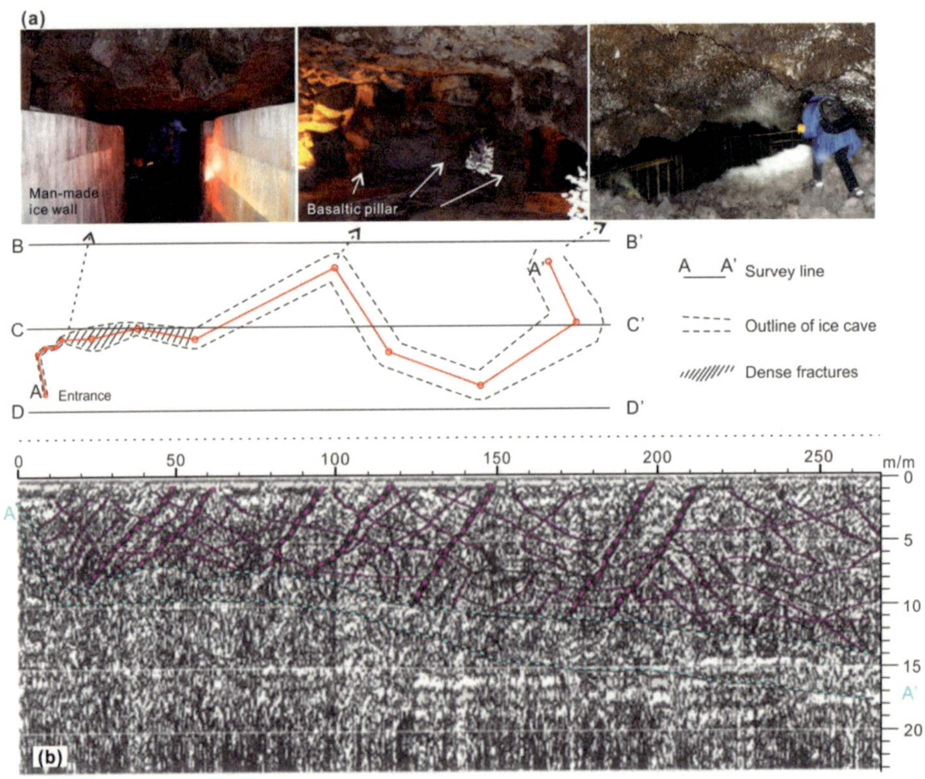

Figure 3.
Top view (a) and geological radar profile (b) of Bailong ice cave at Wudalianchi National Geological Park (modified from [24]).

the entrance of the ice cave. Wang and Zhu measured the geometry and basaltic fracture of Bailong ice cave by applying geological radar (the antenna frequency is 40 MHz) [24] (**Figure 3**). From the radar profile (**Figure 3b**), we can see that seven large fractures with the same dip direction were developed above the ice cave.

Bailong ice cave was not a pure natural ice cave, but ice blocks were moved into the cave to decorate it, and thus its natural conditions were destroyed. The temperature in the ice cave was about −2°C in August 2012. Shuijinggong ice cave is much smaller than Bailong ice cave but with more man-made ice block which was used to decorate it. Therefore, it is difficult to obtain any natural information from Shuijinggong ice cave.

3. Formation of cave

Caves must be developed before the formation of their internal ice. Cave geometry mainly depended on the lithology of their wall rocks. Two types of lithology were involved in these three ice caves: limestone and basalt.

3.1 Limestone as the host rock

The host rock of Ningwu ice cave and Zibaishan ice cave mainly consists of limestone. Limestone is composed of calcium carbonate which is an unsteady material and could be readily dissolved. Limestone areas could become karst landform under wet and warm climate in the geological period. The limestone cave developed predominantly in the perpendicular direction, for example, Ningwu ice cave and Zibaishan ice cave. Generally, the karst region is not suitable for ice preservation. Ice preservation in karst cave may indicate the climate change from warm to cold globally or the presence of incomprehensible cooling mechanisms locally.

3.2 Basalt as the host rock

Basaltic caves (often as tubes) were formed in the process of the viscous lava flow, for example, Wudalianchi ice cave. Mechanical stability of lava tubes depends on their size (diameter), buried depth, and geometry in the cooling period. We conducted a series of two-dimensional thermo-clastic finite element calculations to estimate the effects of these three controlling factors [25].

Figure 4 shows the principle stress distributions around a lava tube of diameter 10 m at the last computing time step. It can be seen from **Figure 4a** that the largest maximum principal stress is at the top and bottom sides of the lava tube, and it is tensile stress. The left and right sides are also tensile stresses, but the magnitude is much smaller. **Figure 4b** is the minimum principal stress distribution. Obviously, the minimum principal stress of the upper and lower sides of the lava tube is tensile stress, while that of the left and right sides is compressive stress, and the magnitude of the stress around the lava tube is equal. These modeling results reveal that the most vulnerable place of the lava tube is on the top and bottom, and the rupture type is tension rupture.

We investigated three controlling factors (diameter, depth, and geometry) of lava tubes in the cooling period by FEM method, and the modeling results were illustrated in **Figure 5**. At the last time step, maximum principal stress of **Figure 5a** (normal lava tube) is smaller than that of **Figure 5b** (bigger lava tube), implying bigger lava tubes were easier ruptured. Similarly, lave tubes with larger buried

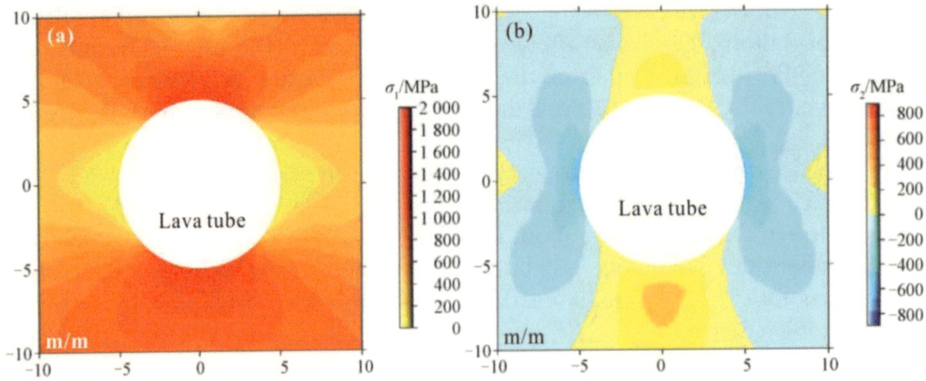

Figure 4.
Principle stress distributions around the lava tube. (a) Maximum principal stress and (b) minimum principal stress (modified from [25]).

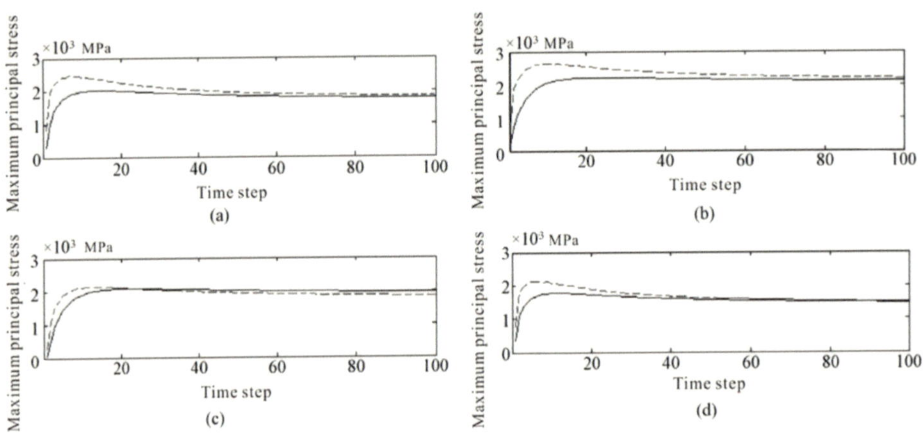

Figure 5.
Maximum principal stresses at the top (dashed line) and bottom (solid line) side of a normal (a), a bigger (b), a deeper (c), and an elliptical (d) lava tube, respectively. The normal lava tube with 10 m diameter and 5 m buried depth. The bigger lava tube with 16 m diameter and 5 m buried depth. The deeper lava tube with 10 m diameter and 8 m buried depth. The elliptical lave tube with 12 m long axis, 8 m short axis, and 5 m buried depth (modified from [25]).

depth would be easier ruptured than normal ones (**Figure 5c**). Elliptical lava tubes would be more stable than normal lava tubes (**Figure 5d**).

4. Formation of ice deposit

After caves were formed, ice may be generated gradually in a suitable climate. For basaltic cave (or lava tube), there may be a clear time boundary between cave formation and ice formation, because lava tubes are steady in water circulation. For limestone cave, there is no such clear time boundary due to limestone instability in wet environments during the geological period.

4.1 Qualitative analysis of cave cooling by air convection

There are variable hypotheses about the preservation mechanism of ice body in Ningwu ice cave. Chen [21] proposed that there is a "cold source" below Ningwu ice cave, which generates the negative geothermal anomaly and then preserves the ice body. Meng et al. [20] ascribed the existence of ice deposit to the combined effect of multiple factors, i.e., geographical location, the "icehouse effect," the "chimney effect," and the "thermal effect" produced by the ice deposit and the "millennial volcano." Unfortunately, they did not supply more details on these factors. Gao et al. [19] considered two aspects, terrain and climate, and proposed that far more cold air than warm air entered the region, and thus the ice cave stayed cold over a year. The subsurface temperature usually rises with depth at a geothermal gradient of 1.0–3.0°C $(100\ m)^{-1}$ [26], which cannot support a permanent "cold source" underground. Furtherly, even if a cold region had somehow formed, it would be warmed up by the geothermal flux through the geological period, because the host rock successively transfer heat energy to the "cold source." At another point, it is a temperate climate outside of the cave. It is difficult to maintain ice deposit in a warm climate without an efficient cooling mechanism. Therefore, we proposed that there must be a sustainable and efficient mechanism to remove the heat from underneath and ensure the existence of the ice deposit.

There is an annual cyclic variation on the air temperature outside Ningwu ice cave: it is warmer than the inside temperature in spring, summer, and autumn, but colder in winter. Because Ningwu ice cave has only a single opening at the top of the cave, cold air of the ice cave could be relatively heavy in spring, summer, and autumn and sinks into the bottom of the cave. Thus, it will not generate air natural thermal convection. In these three seasons, heat energy is transferred by conduction from outside down to the ice cave and from wall rock because of the terrestrial heat flux. Thermal conductivities are low for either wall rock (limestone) or air, and thus the conductive heat transfer efficiency is low. Consequently, the energy conducted to the inside of ice cave in these three seasons is quite limited. However, in winter, it is colder than the inside of the cave, and thus the air outside of ice cave could be heavier than the inside. Gravitational instability is generated, and air thermal convection could occur. The outside cold air promptly flows into the cave to cool it down, and it removes the heat energy out of the cave, which is conducted into the cave from the host rock and through the opening in spring, summer, and autumn. Convective heat transfer is much more efficient than conduction; therefore, the heat energy convected out of the cave in winter is enough to balance the heat conducted into the cave year-round. The annual heat budget of income and output is balanced, so the cave would be in a cyclic state with very small temperature fluctuations, and the average temperature is always lower than 0°C; thus ice deposits in the ice cave can be kept.

Studies in Zibaishan and Wudalianchi ice caves were still at a relatively low level by far. We proposed that similar air convection cooling could occur in these two ice caves.

4.2 Quantitative calculation of ice forming and melting

Accurately, ice forming and melting in caves include at least three physical processes: water flow in porous media (limestone), natural convection of low viscous material (air), and water-ice phase change. The numerical simulation of each process involves complex mathematical method, especially the second physical process. Based on some assumptions, an equivalent method was used to deal with air natural convection [27].

4.2.1 Basic ideas of simulation

Two heat transfer processes must be considered to interpret the existence of ice deposits in Ningwu ice cave, i.e., thermal conduction and convection. The water-ice phase change can reduce the rate of temperature change. So the phase change should be taken into account. The conducting process is governed by conduction equation, which is relatively easy to be computed, while for the convection process, the convection pattern of air and its thermal consequences are hard to determine exactly, because of the complicated geometry of the ice cave and complex varying boundary conditions. In consideration of this, a broadly used, simplified approach is applied in this study: evaluation of Nusselt number (*Nu*) and solving the conductive equation by introducing an equivalent thermal conductivity of the convecting air. For an upright circular tube, the physical relation between the temperature difference of the top and the bottom and *Nu* can be evaluated by adopting the experimental relation of natural convection. The enthalpy method can be adapted to compute the ice-water phase change.

In every computing time step of our modeling process, it is judged whether air convection occurs according to the temperature difference between the top and the bottom of the cave. If there is no convection, the simple conduction problem will be solved, while if convection occurs, an effective conductivity is used in the conduction equation.

4.2.2 Equivalent thermal conductivity

The heat conduction equation and enthalpy method for dealing with water-ice phase change were detailed in our previous work [27]. Equivalent thermal conductivity method was a key point of computing and was shown as follows:

Nu is defined as the ratio of convection heat transfer to pure conduction heat transfer under the same conditions. The efficiency of energy transfer is *Nu* times greater than the conductive efficiency under the same conditions. Therefore, an equivalent thermal conductivity can be introduced, which is *Nu* times greater than the air thermal conductivity [28]. *Nu* is related to physical properties (e.g., viscosity and conductivity of air), to the temperature difference of air at the top and the bottom of the cave, and also to the

geometry of the cave.

Ningwu ice cave can be evaluated by an upright circular tube. For an upright circular tube, *Nu* can be calculated based on experimental fluid thermodynamics studies. Once Eq. (1) is satisfied [29, 30] (the case for Ningwu ice cave), the natural convection heat transfer relation [29, 31] can be expressed as Eq. (2):

$$d/h \geq 35/Gr^{1/4} \quad (1)$$

$$Nu_m = C\,(Gr \cdot Pr)_m^n \quad (2)$$

where d and h in Eqs. (1) and (2) are the diameter and the height of a circular tube respectively; Nu_m is the Nusselt number; Pr is the Prandtl number and is dependent only on the material, e.g., Pr is 0.7 of air; and C and n are constants, the values of which are shown in **Table 1**.

Gr is the Grashof number:

$$Gr = g\beta \Delta T l^3/\upsilon^2 \quad (3)$$

where g is the acceleration of gravity, β is the coefficient of cubical expansion, ΔT is a temperature difference, l is a characteristic length, and υ is the coefficient of kinematic viscosity. The values are $g = 9.8$ m s^{-2}, $\beta = 3.67 \times 10^{-3}$ k^{-1}, $l = 80$ m, and $\upsilon = 13.30 \times 10^{-6}$ m^2 s^{-1} and are substituted into Eq. (3) to obtain

Flow State	Coefficient *C*	Index *n*	*Gr* Application Range
Laminar Flow	0.59	0.25	1.43×10^{4}~3×10^{9}
Transitional Flow	0.0292	0.39	3×10^{9}~2×10^{10}
Turbulent Flow	0.11	0.3333	$>2\times10^{10}$

Table 1.
Gr *number and constant for different flow types [28].*

$$Gr = 1.041 \times 10^{14}\Delta T \tag{4}$$

According to Eq. (4), once the temperature difference is only 10^{-3}°C, *Gr* can reach 1.041×10^{11}. Based on **Table 1**, we infer that natural convection could occur and the fluid state of air is a turbulent flow. The relation between *Nu* to the temperature difference can be obtained when relevant parameters are substituted into Eq. (2):

$$Nu = 11000\,(0.0740\Delta T)^{1/3} \tag{5}$$

In other cases, Eq. (1) may be not satisfied; corresponding experimental physical relations can also be found in this literature [30, 32–33].

Although we took Ningwu ice cave as an example to show the equivalent thermal conductivity method, the method could be used directly to Zibaishan ice cave. Because Zibaishan ice cave is a typical upright circular tube.

4.2.3 Simulated results of ice deposit forming and melting

There is an annually periodic change on the air temperature outside of Ningwu ice cave. Therefore, the air temperature inside the ice cave would show a periodic fluctuation corresponding to the heat conduction and convective processes. The evolution of the air temperature at the bottom of the ice cave is shown in **Figure 6**.

Figure 6a represents the air temperature evolution in Ningwu ice cave during its initial 16 years of the formation process of ice deposit. It can be seen that the air temperature in the cave rises in three seasons (spring, summer, and autumn) and drops in winter only, showing annually periodic variation. The air temperature of Ningwu ice cave drops promptly in winter while rises slowly in other three seasons (spring, summer, and autumn), because the efficiency of heat transfer in these three seasons is much lower than that in winter. If the water-ice phase change process was considered (black line), the dropped rate of air temperature in summer is smaller than that without the water-ice phase change (red line), because the latent heat of the water-ice phase change is required to melt ice near the cave entrance, thus delaying the conduction of heat energy to the bottom of the cave, while the convective cooling in winter is so efficient that the temperature difference is minimized. The temperature decreases below 0°C all year-round after winter cooling for about 5 years. That means the permanent cave ice deposit can be formed and then maintained after winter cooling for about 5 years.

Figure 6b represents the annual cave air temperature periodic fluctuations for the case where the heat transfer process has lasted two centuries, long enough to have evolved to a quasi-stable cyclic state. The amplitude of the cave air temperature fluctuation is about 1.0°C (from −3.9 to −2.9°C). Ningwu ice cave has been

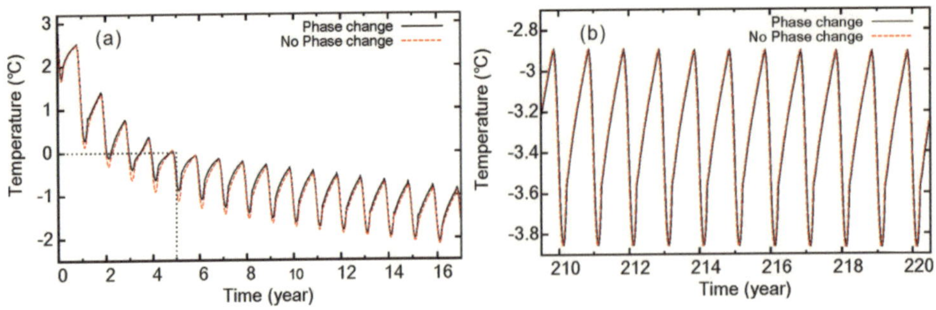

Figure 6.
Ice deposit forming process of Ningwu ice cave. (a) Initial stage of the formation process and (b) quasi-stable stage.

opened to tourists, and about 1500 tourists visited the cave per day from May to October. Consequently, the cave air temperature has been disturbed. The lowest air temperature inside the cave was −1.5°C by our in situ measurement on 5 June 2012. Through the record in the literature, the actual measured inside air of the ice cave ranges between −1.0 [20], −4.0, and −6.0°C [19]. The inconsistent measurements may be contributed to variable measuring methods, measuring time and measuring positions. Similar to what is illustrated in **Figure 6a**, the cave air temperature shows annually periodic fluctuation. The overall rising rate of cave air temperature is lower than its dropping rate, resulting from variable heat transfer efficiency of conduction and convection. The variation on cave air temperature of the model with the water-ice phase change considered (black line) is almost the same as that without the water-ice phase change considered (red line). The reason is that, although the water-ice phase change was considered at the computation, the ice deposit temperature could be always maintained below 0°C when it reaches a quasi-stable cyclic state, and thus no water-ice phase change actually occurred.

The ice deposit in Ningwu ice cave would be melted if there is no air heat convection in winter. Taken the temperature distribution at 220 years (**Figure 6b**) as initial temperature, the evolution of temperature distribution can be computed with or without considering the water-ice phase change effects. The relatively modeling results are presented in **Figure** 7 as a black line and a red line, respectively. They are the same when the air temperature is below the water-ice phase change temperature. However, the ice deposit body takes much longer to be thawed when the water-ice latent heat of melting is taken into account than when it is not considered. It takes 37 years to melt all of ice deposit with the water-ice phase change, while Ningwu ice deposit may be melted completely within 23 years without the water-ice phase change.

4.3 Dynamic balance of water-ice and energy

Water and ice of Ningwu ice cave may be in a dynamic equilibrium state. Ice stalactites and ice stalagmites (**Figure 1a**) can be found everywhere inside of Ningwu ice cave. The observation indicates that water infiltrates into Ningwu ice cave through fractures of limestone throughout the year and then forms ice deposit. Meanwhile, ice deposit at the bottom of Ningwu ice cave could be melted under geothermal flow, and then the produced water infiltrates into places beneath Ningwu ice cave. There are no direct observations to support this inference. This hypothesis may be correct, because of the conservation of water.

The Rayleigh number (*Ra*, a dimensionless number) is related to the buoyancy-driven flow. When *Ra* is below a critical value of that fluid, heat energy is

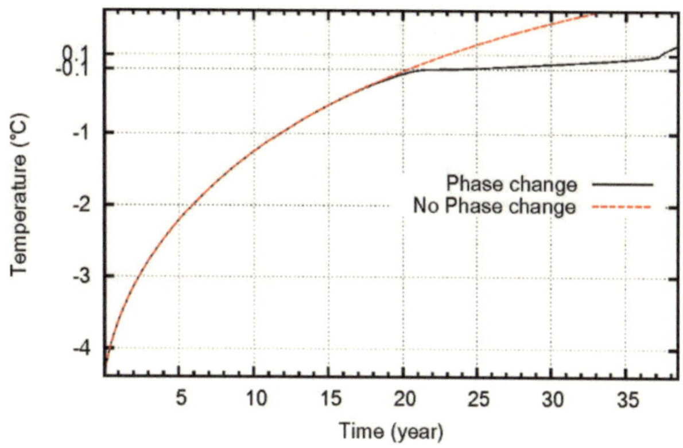

Figure 7.
Illustration of internal temperature evolution of ice cave while ice deposit is melting.

predominantly transferred by conduction; if *Ra* exceeds the critical value, heat energy is mainly transferred by the air natural convection. We proposed that the evolution of ice deposit in Ningwu ice cave is a typical self-regulating process. When too much ice deposit accumulates in the cave, then the cavity will become small gradually. Thus, *Ra* and *Nu* will be reduced, meaning the freezing efficiency decreases, resulting in some of the cave ice deposit melting. Ice deposit melting will lead the cavity to become large, and thus the corresponding *Ra* and *Nu* will be risen. This implies that the freezing efficiency will increase and ice deposit in Ningwu ice cave will grow again.

We proposed that Zibaishan ice cave may be also a self-regulating system, controlled by air natural convection between inside and outside of the ice cave.

5. Discussion

In spring, summer, and autumn, heat energy was transferred into Ningwu ice cave through air and wall rock by thermal conduction, resulting in increasing the air temperature of the ice cave limitedly, because the efficiency of conduction is relatively low. But in winter, the air temperature of the ice cave decreases promptly, caused by air natural thermal convection. The water-ice phase change buffers the energy exchange. Considering these physical mechanisms, the modeling results present that (1) starting from a normal ground temperature distribution, a year-round ice deposit will be produced in the cave within a decade, about 5 years (**Figure 6a**), and the air temperature of the cave will drop gradually for more than a century, and also that (2) the air temperature of the ice cave will finally reach a quasi-stable cyclic state and will vary within a certain range (less than 1.0°C, from −3.9 to −2.9°C). At this stage, the annual total heat energy conducted into the cave and the heat energy removed from the cave by air natural convection are balanced.

Installing an airtight door at a cave entrance is what the Wudalianchi National Geological Park has done to "prevent" the ice deposit from melting every night during the tourist season and the entire winter. When the cave airtight door is closed, in fact, the air convection in winter is blocked. Consequently, cold air cannot enter the cave and cannot remove heat energy from the cave. Accumulation of heat energy conducted by air and rock will eventually cause the ice deposit to melt in Wudalianchi cave. Our modeling results present that it takes less than 40 years to completely melt the whole ice deposit in the cave. This means that Ningwu ice cave is probably not undergoing melting of the ice deposit at present. This study also suggests that scientific management is significant for sustainable application of natural resources. Otherwise, well-meaning acts such as installing a door to completely seal the entrance will actually destroy the ice cave in a few decades.

We expect that no airtight door will be fixed at Zibaishan ice cave entrance and few people enter the ice cave before comprehensive and detailed studies, avoiding affecting its natural conditions.

6. Conclusion

This chapter has described three famous ice caves in China, i.e., Ningwu ice cave, Zibaishan ice cave, and Wudalianchi ice cave. We reviewed quantitative analysis of the formation and preservation mechanism of an ice deposit in Ningwu ice cave, a quasi-static ice cave. The systematic FEM computing leads to the conclusion below: the air natural convection in winter is the crucial controlling factor for forming and sustaining the ice deposit in the cave and can cool the ice cave efficiently. Heat transfer by conduction in spring, summer, and autumn is very limited to warm up the cave. Water-ice phase change plays an important role in summer and can prevent melting of ice. We proposed that Ningwu ice cave and Zibaishan ice cave may be a self-regulating system, respectively, controlled by air natural convection between inside and outside of the ice caves. We expect that no airtight door will be fixed at each ice cave entrance and few people enter the ice cave before comprehensive and detailed studies, avoiding further affecting its natural conditions.

Acknowledgements

We appreciate Prof. Lada Bozic's invitation. Many thanks to Prof. Huai Zhang, Yuanze Zhou, and Guiping Zhao for their kind help in field work. We thank Jiaqi Sun who supplied the pictures of Zibaishan ice cave. This research is supported by the National Natural Science Foundation of China (NSFC), project 41590860 and U1839207.

Author details

Shao-Hua Yang[1] and Yao-Lin Shi[2*]

1 Key Laboratory of Deep-Earth Dynamics, Institute of Geology, Chinese Academy of Geological Sciences, Beijing, China

2 Key Laboratory of Computational Geodynamics of Chinese Academy of Sciences, University of Chinese Academy of Sciences, Beijing, China

*Address all correspondence to: shiyl@ucas.ac.cn

References

[1] May B, Spötl C, Wagenbach D, Dublyansky Y, Liebl J. First investigations of an ice core from Eisriesenwelt cave (Austria). The Cryosphere. 2011;**5**:81-93. DOI: 10.5194/tc-5-81-2011

[2] Obleitner F, Spötl C. The mass and energy balance of ice within the Eisriesenwelt cave, Austria. The Cryosphere. 2011;**5**:245-257. DOI: 10.5194/tc-5-245-2011

[3] Schöner W, Weyss G, Mursch-Radlgruber E. Linkage of cave-ice changes to weather patterns inside and outside the cave Eisriesenwelt (Tennengebirge, Austria). The Cryosphere. 2011;**5**:603-616. DOI: 10.5194/tc-5-603-2011

[4] Bella P. Morphology of ice surface in the Dobšiná ice cave. In: 2nd International Workshop on Ice Caves, Demänovská Dolina, Slovak Republic. 2006

[5] Lalkovič M. On the problems of the ice filling in the Dobšina ice cave. Acta Carsol. 1995;**24**:313-322

[6] Holmlund P, Onac BP, Hansson M, Holmgren K, Mörth M, Nyman M, et al. Assessing the palaeoclimate potential of cave glaciers: The example of the Scărişoara ice cave (Romania). Geografiska Annaler Series A. 2005;**87**:193-201. DOI: 10.1111/j.0435-3676.2005.00252.x

[7] Persoiu A, Onac BP, Wynn JG, Bojar AV, Holmgren K. Stable isotope behavior during cave ice formation by water freezing in Scărişoara ice cave, Romania. Journal of Geophysical Research. 2011;**116**:D02111. DOI: 10.1029/2010JD014477

[8] Luetscher M, Bolius D, Schwikowski M, Schotterer U, Smart PL. Comparison of techniques for dating of subsurface ice from Monlesi ice cave, Switzerland. Journal of Glaciology. 2007;**53**:374-384. DOI: 10.3189/002214307783258503

[9] Luetscher M, Lismonde B, Jeannin PY. Heat exchanges in the heterothermic zone of a karst system: Monlesi cave, Swiss Jura Mountains. Journal of Geophysical Research. 2008;**113**:F02025. DOI: 10.1029/2007JF000892

[10] Peters KF. Geologische und mineralogische Studien aus dem südöstlichen Ungarn, insbesondere aus der Umgegend von Rézbánya. Wien: KK Hof-und Staatsdruckerei; 1861

[11] Laursen L. Climate scientists shine light on cave ice. Science. 2010;**329**:746-747. DOI: 10.1126/science.329.5993.746

[12] Roberts MS, Smart PL, Baker A. Annual trace element variations in a Holocene speleothem. Earth and Planetary Science Letters. 1998;**154**:237-246. DOI: 10.1016/S0012-821X(97)00116-7

[13] Lacelle D, Lauriol B, Clark ID. Seasonal isotopic imprint in moonmilk from Caverne de l'Ours (Quebec, Canada): Implications for climatic reconstruction. Canadian Journal of Earth Sciences. 2004;**41**:1411-1423. DOI: 10.1139/e04-080

[14] Johnson KR, Hu C, Belshaw NS, Henderson GM. Seasonal trace-element and stable-isotope variations in a Chinese speleothem: The potential for high-resolution paleomonsoon reconstruction. Earth and Planetary Science Letters. 2006;**244**:394-407

[15] de Freitas C, Littlejohn R. Cave climate: Assessment of heat and moisture exchange. Journal of Climatology. 1987;**7**:553-569. DOI: 10.1002/joc.3370070604

[16] de Freitas C, Littlejohn R, Clarkson T, Kristament I. Cave climate: Assessment of airflow and ventilation. International Journal of Climatology. 1982;**2**:383-397. DOI: 10.1002/joc.3370020408

[17] Reddy JN, Gartling DK. The Finite Element Method in Heat Transfer and Fluid Dynamics. 3rd ed. Boca Raton: CRC Press; 2010

[18] Gerya T. Introduction to Numerical Geodynamical Modelling. New York: Cambridge University Press; 2010

[19] Gao L, Wang X, Wan X. Analysis of ice cave formation in Ningwu Shanxi. Journal of Taiyuan University of Technology. 2005;**36**:455-458. DOI: 10.3969/j.issn.1007-9432.2005.04.022

[20] Meng X, Zhu D, Shao Z, Yu J, Han J, Meng Q. A discussion on the formation mechanism of the "ten-thousand year-old ice cave" in Shanxi Province. Acta Geoscientia Sinica. 2006;**27**:163-168. DOI: 10.3321/j.issn:1006-3021.2006.02.011

[21] Chen S. Cave Tourism Science. Fuzhou: Fujian People's Publishing House; 2003

[22] Shao Z, Meng X, Zhu D, Yu J, Han J, Meng Q, et al. Detecation for the spatial distribution of "ten-thousand ice cave" in Ningwu, ShanXi province. Journal of Jilin University (Earth Science Edition). 2007;**37**:961-966. DOI: 10.3969/j.issn.1671-5888.2007.05.019

[23] Shicai C. Three characteristics of Zibaishan ice cave. Chinese National Geography. 2017;(2):134. (in Chinese)

[24] Wang R, Zhu C. An application of geological radar on the fracture survey of Dixiabinghe ice cave at Wudalianchi, Heilongjiang Science and Technology of Water Conservancy. 2012;**40**(4): 268-269. (in Chinese)

[25] Li ZC, Shi YL, Zhang ZY, et al. Calculation of thermal stress during the cooling of lava tubes. Journal of University of Chinese Academy of Sciences. 2016;**33**(3):412-420

[26] Hu S, He L, Wang J. Compilation of heat flow data in the China continental area (3rd ed.). Chinese Journal of Geophysics. 2001;**44**:611-626. DOI: 10.3321/j.issn:0001-5733.2001.05.005

[27] Yang SH, Shi YL. Numerical simulation of formation and preservation of Ningwu ice cave, Shanxi, China. The Cryosphere. 2015;**9**:1983-1993

[28] Schmeling H, Marquart G. A scaling law for approximating porous hydrothermal convection by an equivalent thermal conductivity: Theory and application to the cooling oceanic lithosphere. Geophysical Journal International. 2014;**197**:645-664. DOI: 10.1093/gji/ggu022

[29] Sparrow E, Gregg J. Laminar free convection heat transfer from the outer surface of a vertical circular cylinder. Transactions ASME. 1956;**78**:1823-1829

[30] Yang S, Tao W. Heat Transfer. 4th ed. Beijing: Higher Education Press; 2006

[31] Incropera FP, Lavine AS, DeWitt DP. Fundamentals of Heat and Mass Transfer. 11th ed. Hoboken, NJ: John Wiley & Sons Incorporated; 2011

[32] Cebeci T. Laminar-free-convective-heat transfer from the outer surface of a vertical slender circular cylinder. In: Proceedings of the 5th International Conference; Tokyo, Japan. 1974. pp. 15-19

[33] Minkowycz W, Sparrow E. Local nonsimilar solutions for natural convection on a vertical cylinder. Journal of Heat Transfer. 1974;**96**:178-183. DOI: 10.1115/1.3450161